Gene Expression - From Code to Cure

Edited by Preeti Dabas,
Rafael Trindade Maia, Camila Oliveira,
Glaucia V. Faheina-Martins
and Magnólia De Araújo Campos

Published in London, United Kingdom

Gene Expression - From Code to Cure
http://dx.doi.org/10.5772/intechopen.1008065
Edited by Preeti Dabas, Rafael Trindade Maia, Camila Oliveira, Glaucia V. Faheina-Martins and Magnólia De Araújo Campos

Contributors
Akula Sindhu Bhargavi, Anant Jaiswal, Carol Bernstein, Eric Deneault, Harris Bernstein, Jiabin Duan, Md Abdur Razzaque, Prakash Sulur Muthusamy, Shaohua Qi, Tao Wang, Timothy Osborne, Vasanthakumar Karumannan, Xian Hong, Yan-Ji Lu

First published in London, United Kingdom, 2026 by IntechOpen
IntechOpen is the global imprint of INTECHOPEN LIMITED, registered in England and Wales, registration number: 11086078, 167-169 Great Portland Street, London, W1W 5PF, United Kingdom

For EU product safety concerns: IN TECH d.o.o., Prolaz Marije Krucifikse Kozulić 3, 51000 Rijeka, Croatia, info@intechopen.com or visit our website at intechopen.com.

British Library Cataloguing-in-Publication Data
A catalogue record for this book is available from the British Library

Gene Expression - From Code to Cure
Edited by Preeti Dabas, Rafael Trindade Maia, Camila Oliveira, Glaucia V. Faheina-Martins and Magnólia De Araújo Campos
p. cm.

This title is part of the Genetics Book Series, Volume 14
Topic: Genomics
Series Editor: Kenji Ikehara
Topic Editor: Irina Vlasova-St. Louis

Print ISBN 978-1-80631-052-4
Online ISBN 978-1-80631-051-7
eBook (PDF) ISBN 978-1-80631-053-1
ISSN 3049-7094

IntechOpen Book Series
Genetics
Volume 14

Aims and Scope of the Series

"Genetics," which has been proud of its tradition since Mendel presented his research results in 1865, initially progressed quite slowly due to simple observational approaches of individuals and groups. However, the discovery of double-stranded DNA by Watson and Crick about 70 years ago triggered rapid progress in life sciences, including genetics, which was primarily conducted using Escherichia coli and bacteriophages infecting E. coli. Subsequently, genetics has achieved remarkable developments, such as understanding genetic disorders, including cancers, through research on the biogenesis and differentiation of plants and animals.
The two topics of this book series - Human Genetics, and Genomics - will address important areas of advancement in genetics.
Human Genetics: After fundamental genetics, initially studied with the main goal of revealing the functions of individual genes and proteins, genetics expanded from understanding the genetic system itself to understanding many infectious diseases caused by bacteria and viruses. Consequently, human beings are now overcoming infectious diseases by developing medicinal chemicals, including antibiotics and vaccines. However, genetic disorders remain challenging to cure up to now. Nevertheless, even the cure for them, including various cancers, is coming closer to reality due to the rapid progress of human genetics. In this way, the welfare of human life continues to improve, and even longevity, which was once a dream, has been achieved to some extent in recent years.
Genomics: On the other hand, the understanding of the comprehensive interrelationship of whole genes or whole proteins functioning in one organism has become possible now, as research has entered the era of genomics, owing to the rapid progress of base sequence analysis and bioinformatics. The development of genomics has further made it possible to understand the evolutionary processes of organisms through comparative studies among the genomes of many organisms.
This book series will discuss the findings obtained during the advancement of human genetics and genomics. It is also expected that this series will trigger the formation of a better world composed of human beings and all other organisms on Earth through discussions of research results obtained under the development of general genetics.

Meet the Series Editor

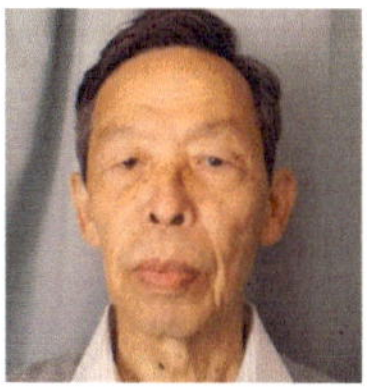

Kenji Ikehara graduated from the Department of Industrial Chemistry, Faculty of Engineering, Kyoto University in 1968. He received his B. Eng. (1968) and subsequently earned M. Eng. (1970) and D. Eng. (1976) degrees from Kyoto University. He began his career as a research associate in the Faculty of Science at the University of Tokyo before moving on to become an associate professor in the Faculty of Science at Nara Women's University. He was later promoted to professor and subsequently served as the dean of the Faculty of Science at Nara Women's University. Additionally, he held the position of director at the Nara Study Center of the Open University of Japan. For approximately 15 years, he focused his research on sporulation initiation of Bacillus subtilis. Later, he shifted his focus to the origins and evolutionary processes of microbial genes, the genetic code, proteins, and life. He has proposed several hypotheses, including the GC-NSF(a) hypothesis on the origin of genes, the GNC-SNS hypothesis on the genetic code, the protein 0th-order structure hypothesis on the origin of proteins, and the [GADV]-protein world hypothesis (GADV hypothesis) on the origin of life. Furthermore, he served as the local chair of the International Conference, Origin 2014, held in Nara in 2014.

Meet the Volume Editors

Dr. Preeti Dabas, a motivated scientific professional, grew up in India and earned her Ph.D. from Guru Gobind Singh Indraprastha University in Delhi, India, where she focused on yeast genetics, molecular biology, biochemistry, and biophysical techniques. She conducted her postdoctoral research at St. Jude Children's Research Hospital in Memphis, USA, specializing in gene expression, signal transduction, DNA repair, epigenetics, stem cell biology, and translational therapeutics, as well as hematopoietic stem cell health in development and disease. Dr. Dabas has authored publications in leading journals, including *Science, Science Translational Medicine, Nucleic Acids Research*, and *IUBMB Life,* and has contributed multiple review articles and book chapters. She also serves as an academic editor and coeditor of international scientific volumes. Her research achievements have been recognized by Marquis Who's Who in America and multiple competitive fellowships.

Rafael Trindade Maia holds a Ph.D. in Animal Biology from the Federal University of Pernambuco (UFPE), a master's degree in Genetics, Conservation and Evolutionary Biology from the National Institute for Amazonian Research (INPA), and a degree in Biological Sciences from the Federal Rural University of Pernambuco (UFRPE). He is currently a professor at the Federal University of Campina Grande (UFCG) and a faculty member of graduate programs in Natural Sciences and Biotechnology (UFCG) and Cellular and Molecular Biology (UFPB). His research focuses on computational biology, molecular modeling, docking, molecular dynamics, genetics, and phylogeny, with strong involvement in science and biology education.

Magnólia A. Campos is a biologist, has a Master's in Agronomy / Plant Breeding from the Federal University of Pelotas, and a Ph.D. in Biological Sciences / Molecular Biology from the University of Brasília (2002). She had 5 years of experience in genomic sciences as a postdoctoral researcher at the Federal University of Lavras/ Agronomic Institute (IAC). Since 2008, she has been a Professor at the Federal University of Campina Grande (UFCG) at the Education and Health Center (CES, Cuité Campus). She is the Coordinator of the Research Group on Biotechnology Applied to the Semi-Arid Region and of the Biotechnology Laboratory at CES. She is a faculty member of the graduate programs in Natural Sciences and Biotechnology (UFCG) and was a recipient of a CNPq research productivity scholarship. She has experience in plant biotechnology, working mainly on the following topics: genomics, bioinformatics, tissue culture and plant cells, plant genetic transformation, gene expression during plant-microbe interactions, and expression of heterologous proteins in bacteria.

Glaucia Veríssimo Faheina Martins holds a Ph.D. in Biotechnology from RENORBIO–Federal University of Paraíba (UFPB), a Master's degree in Bioactive Natural and Synthetic Products, and a Bachelor's degree in Biological Sciences from UFPB. She is an Associate Professor at UFPB and a permanent faculty member of the Graduate Program in Cellular and Molecular Biology. She coordinates the Laboratory of Toxicology and Cancer Genetics (TOXGEN), where her research focuses on molecular genetics and cellular pharmacology, with an emphasis on identifying diagnostic and prognostic molecular biomarkers for colorectal cancer and molecular targets for therapy. She collaborates with multidisciplinary research groups and serves as Head of the Department of Molecular Biology at UFPB.

Camila Franco Batista de Oliveira is a biologist with a master's degree in Neuroscience and a Ph.D. in Bioinformatics from the Federal University of Minas Gerais (UFMG), where she also completed postdoctoral training. Her work focuses on immunology, biotechnology, and bioinformatics, with an emphasis on vaccine development, diagnostic tools, and molecular modeling. She has extensive experience in laboratory and computational techniques applied to human and animal health. She worked at Fiocruz on Polio and COVID-19 diagnostics. She participates in research projects at federal institutions such as IFPB, IFCE, UFCG, and UnB, and serves as faculty in the Graduate Program in Bioinformatics at UFRN. She is the author of articles in high-impact international journals, contributing significantly to immunology, toxinology, and applied bioinformatics.

Contents

Preface

The molecular underpinnings of life are intricate, dynamic, and profoundly influential. From the sequences that encode our genes to the mechanisms that regulate their expression, the "code of life" orchestrates development, health, and disease. Recent decades have brought unprecedented advances in our ability to read, edit, and even rewrite this code, ushering in a new era of molecular medicine. Gene Expression - From Code to Cure is designed to guide readers through this scientific revolution. By integrating gene editing, cellular and RNA-based therapies, DNA repair research, and epigenetic regulation, this book connects fundamental biological principles with cutting-edge therapeutic innovations. Together, these advances are transforming our understanding of disease and reshaping the future of precision medicine.

The first section, *Modern Genetic Therapies*, explores the tools and strategies that enable increasingly precise manipulation of the genetic code and are reshaping modern medicine. It begins with the chapter "Editing the Code of Life", which traces the evolution of gene-editing technologies from early discoveries in DNA biology to the versatility of contemporary CRISPR-based systems. Techniques such as Cas9-mediated editing, base editing, prime editing, and click editing are highlighted for their ability to correct genetic defects both ex vivo and in vivo, offering new possibilities for treating monogenic and complex diseases, while also addressing ongoing challenges such as off-target effects. The second chapter, "Gene, Cell, and RNA Therapies and Its Recent Advances: An Overview", examines how these molecular tools are translated into therapeutic applications. It surveys the rapidly expanding landscape of genetic, cellular, and RNA-based interventions in development and clinical practice, emphasizing recent advances and their potential to treat previously intractable diseases. Together, these chapters set the foundation for understanding how molecular-level interventions can be transformed into life-changing therapies.

Maintaining the integrity of the genome is central to health and longevity, yet DNA is constantly under assault from environmental and endogenous stresses. The second section, *DNA Integrity, Repair, and Aging*, delves into the molecular mechanisms that safeguard DNA and the consequences when these systems fail. The chapter "DNA Damage, DNA Repair, and Epigenetic Alterations are Central to Aging", examines how unrepaired DNA damage and epigenetic modifications contribute to aging at the cellular level. Evidence from mammalian studies and human genetic disorders highlights the interplay between DNA repair processes and lasting epigenetic changes that influence gene expression, ultimately shaping the aging phenotype. The second chapter in this section, "The Art of Imperfect Joining: NHEJ in DNA Repair and Disease," explores nonhomologous end joining (NHEJ), the primary pathway for repairing DNA double-strand breaks in mammals. This chapter details the molecular machinery of NHEJ, its regulation, and its dual role as both a genome protector and a potential mutagen. It further discusses how NHEJ dysfunction contributes to diseases such as cancer, immunodeficiency, and neurodegeneration, and reviews therapeutic strategies targeting this pathway. Together, these chapters illuminate how understanding DNA repair and its imperfections provides a foundation for precision medicine.

While the genetic code provides the blueprint, epigenetic regulation defines how that blueprint is interpreted. The final section, *Epigenetics and Inheritance: Mechanisms and Metabolic Regulation*, examines how gene expression is controlled beyond the DNA sequence itself through the dynamic interplay between gene expression, chromatin modifications, and metabolic signals. The first chapter in this section, "Epigenetic Inheritance: Maintaining and Fine-Tuning of Origin-Specific Features Across Generations", explores the mechanisms by which epigenetic information is propagated, maintained, and adapted across cell divisions and generations, revealing how life preserves identity while remaining flexible. The second chapter, "Key Role for Histone Modifiers in Coupling Metabolism to Epigenetic Reprogramming Associated with 'Trained Immunity in Innate Immune Cells", highlights how metabolic cues intersect with chromatin regulation to modulate immune memory and responsiveness. This emerging understanding underscores that gene expression is not static but is dynamically regulated by both intrinsic and extrinsic signals.

Across all sections, this book highlights the close interplay between gene expression, DNA repair, epigenetic regulation, and therapeutic innovation. By integrating these fields, this book, Gene Expression - From Code to Cure, provides a roadmap for researchers, clinicians, and students eager to explore how molecular insights can be translated into transformative medical interventions. I hope that this book will serve as both a valuable resource and a source of inspiration, providing a window into a future where understanding the code of life leads to cures once thought impossible.

I am deeply grateful to my coeditors, Prof. Rafael Trindade Maia, Dr. Glaucia Verissimo Faheina Martins, Dr. Magnólia de Araújo Campos, and Dr. Camila Franco, as well as all contributing authors, whose expertise, dedication, and collaborative spirit were essential to the completion of this project. Their thoughtful contributions and commitment to scientific excellence have greatly enriched this volume. I also thank everyone involved in the editorial and production process for their support and professionalism. This book is truly a collective achievement, and I am honored to have worked with such an outstanding group of colleagues.

Preeti Dabas
St. Jude Children's Research Hospital,
Memphis, USA

Rafael Trindade Maia
Universidade Federal de Campina Grande,
Campina Grande, Brazil

Camila Oliveira
Instituto Federal de Educação Ciência e
Tecnologia da Paraíba,
Paraíba, Brazil

Glaucia V. Faheina-Martins
Universidade Federal da Paraíba,
João Pessoa, Brazil

Magnólia De Araújo Campos
Universidade Federal de Campina Grande,
Campina Grande, Brazil

Section 1

Modern Genetic Therapies

Chapter 1

Editing the Code of Life

Eric Deneault

Abstract

From the discovery of DNA to the development of highly precise and versatile molecular tools to modify it, gene editing has evolved to radically transform how we fight cancer and genetic diseases. This chapter describes key milestones that led to CRISPR-based technologies becoming specific tools for genome editing. Techniques based on Cas9 nuclease, base editing, prime editing, and click editing now enable accurate ex vivo and in vivo gene corrections. The expansion to other CRISPR orthologs allows for more diverse genetic modifications. These breakthroughs offer hope for addressing both monogenic and complex disorders. Despite constant progress, ongoing research is required to manage off-target effects and collateral damage that may co-occur during the gene-editing process.

Keywords: DNA, gene editing, CRISPR-Cas9, cancer, genetic disorders

1. Introduction

In the 1860s, a new substance was extracted from the nuclei of human leukocytes by the young scientist Johann Friedrich Miescher, showing distinct properties compared to all types of protein and other organic molecules known at the time [1]. For example, it exhibited high levels of organically bound phosphorus and resisted digestion by the proteolytic enzymes secreted by the pig stomach. Miescher came close to determining the precise function of this mysterious molecule he named "nuclein." However, it took another ~ 75 years to prove that this molecule, eventually renamed deoxyribonucleic acid (DNA), was the carrier of hereditary traits in living organisms [2].

After the DNA double helix structure was solved in 1953 [3] and the genetic code cracked during the following years, scientists began to envision the possibility of intentionally and precisely modifying the DNA sequence. Although the tools for efficient gene editing were not yet available, the dream of correcting pathogenic mutations, silencing deleterious genes, or enhancing beneficial traits had already taken root in the scientific imagination.

Editing the DNA sequence is important to cure genetic diseases by correcting the responsible root mutations, and not only alleviating symptoms with drugs. This offers durable therapies with a single treatment. For example, sickle cell disease (SCD) and beta-thalassemia can be treated with Casgevy™, the first approved gene-editing therapy, by modifying the DNA sequence of a specific enhancer in patients' blood stem cells [4].

2. Tracing original tools to edit the code of life

Over the following decades, this vision became a reality with the development of targeted genome engineering technologies. Restriction enzymes were the first tools to enable precise DNA manipulation by cutting at specific recognition sequences, opening the door to recombinant DNA technology [5] (**Figure 1**). This innovation allowed the transfer of genetic material between different organisms, but the nonprogrammable nature of most restriction enzymes limited their utility in targeting specific loci.

To address the issue of programmability, scientists developed engineered nucleases. Zinc finger nucleases (ZFNs) combine the DNA cleavage domain of the FokI restriction enzyme with customizable zinc finger protein modules, each recognizing specific DNA base triplets [6] (**Figure 1**). These modules collectively guide nuclease specificity. ZFNs later enabled targeted double-strand breaks (DSBs) and the successful correction of a mutation causing severe combined immunodeficiency (SCID) [7]. However, the complexity of assembling zinc finger arrays for each new target limited their practicality.

A few years later, transcription activator-like effector nucleases (TALENs) were shown to be easier to design due to the more flexible nature of their modular repeat structures [8, 9] (**Figure 1**). TALENs are less complex to design and clone compared to the rigid zinc finger motifs used in ZFNs. Each TALEN module recognizes a single DNA base, which facilitates the assembly, whereas each ZFN module recognizes overlapping triplets, which are more complex to engineer. TALENs also exhibit lower off-target effects. Yet, both ZFNs and TALENs required time-consuming protein engineering for each new gene target, which proved impractical for broad implementation.

3. Discovery of a simple and versatile editing system

In 1987, the first observation of clustered direct DNA repeats in bacteria was published, though their function remained unclear [10]. Subsequent studies revealed these repeats as part of a bacterial adaptive immune system against viruses. This system was named clustered regularly interspaced short palindromic repeats (CRISPR) in 2002 [11]. The short repeats were shown to be interspaced with fragments derived from invading bacteriophages or plasmids [12]. When the same

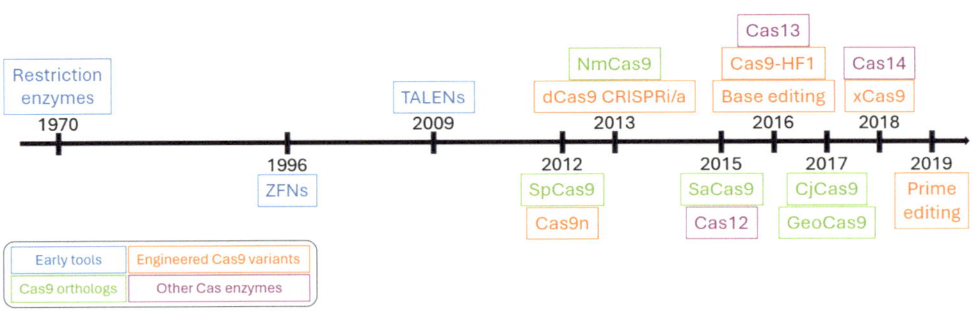

Figure 1.
Timeline of gene-editing tool evolution.

invader reappears, the CRISPR array is transcribed, and CRISPR-associated (Cas) proteins process the resulting RNA into CRISPR RNAs (crRNAs), which direct Cas nucleases to cleave matching foreign DNA.

Among them, Cas9 has been the most widely applied in mammalian gene editing. The type II CRISPR-Cas9 system from the bacterium *Streptococcus pyogenes*, first described in 2012, uses a guide RNA (gRNA) made by fusing a crRNA and a trans-activating CRISPR RNA (tracrRNA) [13] (**Figure 1**). The crRNA contains a 20-nucleotide spacer complementary to the DNA target, while the tracrRNA stabilizes the complex with DNA. Together, they form the single-guide RNA (sgRNA), which directs Cas9 to a target sequence next to a protospacer adjacent motif (PAM), typically made of the nucleotides NGG (N = any nucleotide; G = guanine) [13]. The PAM is not directly involved in the guidance of Cas9 and does not interact directly with the crRNA, but it is required for Cas9 to form the RNA–DNA duplex [14]. Cas9 then makes a DSB at the target site using two internal nuclease domains, HNH and RuvC, without requiring dimerization [13]. The short sgRNA is easily customizable and synthesizable, eliminating the need for engineered proteins, unlike ZFNs or TALENs. The simplicity, efficiency, and adaptability of CRISPR-Cas9 have revolutionized gene editing, enabling applications in a wide range of organisms, including humans [15–17].

4. Large repertoire of Cas9 enzymes

In recent years, dozens of Cas9 orthologs have been characterized in diverse bacterial species. These Cas9 versions differ in PAM specificity (with at least 50 different PAM sequence requirements identified so far), size, editing efficiency, suitability for delivery in therapeutic contexts, temperature dependence, and staggered-end DNA target cleavage [18]. Among the most studied and widely used orthologs, SpCas9 (from *Streptococcus pyogenes*), NmCas9 (*Neisseria meningitidis*), and SaCas9 (*Staphylococcus aureus*) require a different PAM and are smaller than SpCas9 [19, 20] (**Figure 1**). Therefore, they are more amenable to recombinant adeno-associated virus (rAAV) vector all-in-one delivery, which can only package ~4.7 kb of transgene cargo (**Table 1**). Moreover, CjCas9

Cas9 ortholog	PAM	Size (amino acids)	Species	Characteristics	Reference
SpCas9	NGG	1,368	*Streptococcus pyogenes*	Widely used, broad applications	[13]
NmCas9	NNNNGATT	1,083	*Neisseria meningitidis*	Smaller than SpCas9	[19]
SaCas9	NNGRRT	1,053	*Staphylococcus aureus*	Smaller than SpCas9, AAV-packagable	[20]
CjCas9	NNNNACAC or NNNNRYAC	984	*Campylobacter jejuni*	Smallest Cas9, AAV-packagable	[21]
GeoCas9	NNNNCRAA	1,087	*Geobacillus stearothermophilus*	Active up to 70 °C	[22]

Table 1.
Most studied Cas9 orthologs presenting different characteristics.

SpCas9 variant	PAM	Characteristics	Reference
Cas9n	NGG	Cuts one strand	[13, 15, 17]
Cas9n-BE	NGG	Deamination of nucleotides	[30, 31]
Cas9n-PE		All small modifications	[35–37]
dCas9-CRISPRi		Repression of transcription	[38]
dCas9-CRISPRa		Activation of transcription	[39]
Cas9-HF1		High fidelity, low off-target	[40]
HiFi SpCas9			[41]
eSpCas9			[42]
HypaCas9			[43]
xCas9	NG, GAA, GAT	Broader PAM compatibility	[44]
SpCas9-NG	NG		[45]
SpRY	Near-PAMless	Broadest PAM compatibility	[46]

Table 2.
Most used engineered Cas9 variants with improved functions.

(*Campylobacter jejuni*) is one of the smallest Cas9 orthologs identified so far and is highly specific [21], while GeoCas9 (*Geobacillus stearothermophilus*) can tolerate high functional temperatures [22] (**Figure 1**; **Table 1**).

In addition, a series of SpCas9 variants have been engineered to gain or improve various functions. For example, point mutations have been introduced in catalytic residues of the two nuclease domains of Cas9, that is, H840A in HNH, or D10A in RuvC, to convert Cas9 into a nickase (Cas9n) that cuts only one strand (**Figure 1**; **Table 2**) [13, 15, 17]. To generate a DSB, two close offset sgRNAs targeting opposite strands are needed, increasing target specificity. Since single-strand nicks are typically repaired with high fidelity, this dual-nickase strategy reduces off-target effects. For instance, Cas9n-D10A has proven effective in minimizing off-target small insertions and deletions (indels) in various cells and gene editing experiments [23–29].

Base editing (BE) involves the fusion of Cas9n (D10A) to a deaminase enzyme (**Figure 1**; **Table 2**), enabling the conversion of one DNA base pair into another without creating any DSB. Cytosine base editors (CBEs) mediate C–to–T transitions [30], adenine base editors (ABEs) enable A–to–G changes [31], and newer variants can perform multiple base transitions. By avoiding DSBs, BE reduces the risk of indels. BE has shown strong potential for correcting numerous pathogenic mutations in vivo. However, challenges remain since deaminases can also affect other nucleotides within a specific window, which can lead to unintended "bystander" edits nearby. Off-target DNA and RNA deamination have been reported, particularly with CBEs [32, 33]. However, recent phage-assisted engineering in BE has created APOBEC and TadA variants that minimize off-target DNA and RNA editing activity, while maintaining efficient on-target DNA BE, for example, ABE8e [34].

Prime editing (PE) is a precise gene editing technology that enables all 12 base substitutions, in addition to small insertions (≤40 bp), and deletions (≤80 bp) without inducing DSBs nor bystander mutations (**Figure 1**). It fuses Cas9n (H840A) to a reverse transcriptase (RT), guided by a prime editing guide RNA (pegRNA) that encodes both the target sequence and the intended edit (**Table 2**) [35]. PE efficiency highly depends on optimizing the reverse transcription template (RTT) and primer binding site (PBS) for each target. New variants such as PE3 and PE5 enhance efficiency but introduce a second nick to favor edit incorporation, risking indels [36]. PEmax, PE6 and PE7 further improve efficiency through engineered components and small peptide fusion [36, 37, 47].

BE and PE are both precise genome editing tools that avoid DSBs, but they differ in mechanism, efficiency, and safety profile. PE is generally less efficient that BE due to a more complex system that must initiate nicking, reverse transcription, flap resolution, and repair, requiring many coordinated cellular responses. In contrast, BE uses a simpler fusion of Cas9n and a deaminase, which acts directly on the DNA base without requiring synthesis of a new DNA strand. In addition, PE is considered safer than BE since it requires both sequence recognition by the pegRNA and successful RT-mediated DNA synthesis, making unintended edits less likely. Due to their large size, in vivo delivery of base and prime editors remains a major challenge. AAVs often require dual-vector strategies due to their small packaging capacity. Nonviral approaches such as lipid nanoparticles (LNPs) only offer transient expression. However, emerging delivery systems are being developed, like virus-like particles (VLPs) or polymer-based nanocarriers, to improve delivery efficiency and safety.

The catalytic activity of Cas9 was completely abrogated by introducing both H840A and D10A mutations, thereby generating dead Cas9 (dCas9), which still retained its DNA-binding ability [13]. Since dCas9 can block transcription through steric hindrance, this tool was named CRISPR interference (CRISPRi) [38] (**Figure 1**; **Table 2**). To expand this opportunity, various effector proteins have been fused to dCas9 to modulate gene expression without altering the DNA sequence. For example, epigenetic effectors targeting DNA methylation or histone modifications can regulate gene expression, offering therapeutic potential for epigenetically driven diseases. Repressors like KRAB have been used for CRISPRi, while activators like VP64 can drive transcriptional activation (CRISPRa) [39] (**Table 2**).

Extensive engineering of SpCas9 has led to impressive specificity and less PAM restriction. For example, specific mutations were introduced and gave rise to Cas9-HF1 [40] (**Figure 1**), HiFi SpCas9 [41], eSpCas9 [42] and HypaCas9 [43], reducing nontarget DNA binding and minimizing off-target effects (**Table 2**). Targetable PAMs were also expanded to NG, GAA, and GAT with xCas9 [44] (**Figure 1**), while maintaining high specificity (**Table 2**). Elimination of the requirement for the third position of PAM allowed targeting of NG PAMs rather than NGG, thereby increasing target range of SpCas9-NG [45] (**Table 2**). Furthermore, SpRY [46] was developed with a highly flexible PAM recognition, targeting NRN and NYN, which is almost PAM-free (**Table 2**).

5. Additional Cas enzymes for gene editing

Beyond Cas9, other CRISPR systems have expanded the gene-editing toolkit. For example, type V CRISPR-Cas12 targets double-stranded DNA (dsDNA) with a TTN PAM and creates staggered DSBs with 4–5 nucleotides 5′ overhangs, without tracrRNA [48] (**Figure 1**). This can be advantageous for precise ligation. Cas12 also presents nonspecific single-stranded DNA (ssDNA) collateral cleavage properties, which can be used for the rapid diagnosis of infectious diseases through sensitive and cost-effective DNA diagnostics [49].

Type VI CRISPR-Cas13 is unique for targeting single-stranded RNA (ssRNA) instead of DNA and enables programmable mRNA interference [50] (**Figure 1**). RNA editing circumvents permanent genomic changes and DNA off-target risks. Cas13 uses a guide crRNA to target and cleave specific mRNAs, effectively knocking down gene expression. Additionally, catalytically inactive dCas13 can be fused to adenosine deaminases acting on RNA (ADAR) to convert adenosine (A) to inosine (I), which is read as guanine [51]. Similarly, Cas13-based systems like ceRBE enable programmable A–to–I and C–to–U conversions without cleaving RNA. However, mRNA has a short life and is continually transcribed; thus, sustained expression of RNA editors is required for lasting effects, which may increase the risk of bystander or unintended genomic alterations. Due to its nonspecific ssRNA collateral cleavage capacity, Cas13 is also used for rapid RNA virus diagnostics with extreme sensitivity and single-base mismatch specificity [52].

Type V–F CRISPR-Cas14 specifically targets ssDNA and is ultra-small (~400–700 amino acids), which makes it ideal for packaging in AAV [53] (**Figure 1**). Cas14 uses a single crRNA, so no tracrRNA is needed, and no PAM is required. Once activated by binding to a target ssDNA, Cas14 can also cleave other nearby ssDNA, which is useful in DNA diagnostics. However, its specific ssDNA activity makes Cas14 unsuitable for editing dsDNA, which is currently not ideal for genome editing.

All of these Cas enzyme forms have broadened CRISPR applications beyond gene editing to transcription regulation and nucleic acid diagnostics.

6. Restoring CRISPR-induced DNA lesions

The sgRNA-Cas9 nuclease complex does not directly execute the genetic edits. Instead, it acts as a guide, directing other components to the targeted locus. The actual genomic modification is carried out by downstream effectors, such as the host cell's DNA repair machinery or enzymes fused to Cas9, like deaminases (for BE) or RT (for PE). DSBs induced by Cas9 are typically repaired by nonhomologous end joining (NHEJ), a fast but error-prone pathway that often introduces indels [15]. If an indel disrupts the coding frame, it can generate a premature termination codon (PTC), triggering nonsense-mediated decay (NMD) and knocking out gene expression [54].

Alternatively, homology-directed repair (HDR) can enable precise gene modifications in the presence of a repair template. This occurs predominantly during the S and G2 phases of the cell cycle. Exogenous templates, such as single-stranded oligodeoxynucleotides (ssODNs) or plasmids, can be used in combination with the sgRNA-Cas9 complex to introduce very precise point mutations or corrections [55–61].

7. Unintended collateral damage

Therapeutic gene editing offers tremendous potential but also raises important safety concerns. For example, the CRISPR-Cas9 system can tolerate a few

Approach	Target or strategy	Editing tool	Outcome	Notes	Reference
Direct editing of tumor cells	Oncogene inactivation (MYC) or tumor suppressor reactivation (TP53)	Cas9 nuclease	Proof-of-concept editing of cancer driver mutations	Editing efficiency is often suboptimal, leaving residual tumor cells	[64]
Antigen targeting (CAR T- and NK-cells)	CD19 (B-cell malignancies), BCMA (multiple myeloma)	Retro-/lenti-virus or rAAV KI	Durable tumor killing	High costs, concatemer insertions risk	[65–67]
Nonviral KI (HDR)	CAR insertion with long ssDNA flanked by dsDNA Cas9 sites (ssCTS)	HiFi SpCas9 nuclease	Comparable efficiency to rAAV, reduced toxicity, and scalable manufacturing	Avoids rAAV drawbacks	[68]
2-in-1 KO/KI	TRAC KO + CAR insertion at the TRAC locus	Cas9 nuclease	Prevents GvHD, increases persistence	Dual function: KO + controlled CAR expression	[67]
Ortholog KI and ABE KO	CAR KI + B2M and CIITA KOs	AsCas12a and SpCas9-ABE	Reduced simultaneous DSBs	Fewer translocations	[69]
CAR T multiplex KO	B2M, CIITA, PDCD1, CD3E, PVR	ABE8e mRNA	85–95% KO efficiency, better persistence, and antitumor efficacy	Reduced DSBs = fewer translocations	[70]
CAR NK multiplex KO	AHR, CISH, KLRG1, TIGIT, PDCD1	ABE8e mRNA	65–100% KO, improved CAR-NK function	More resilience against TME	[71]
Other functional KOs	CD38 prevents fratricide, CALHM2 improves cytotoxicity, and CREM enhances resistance to immunosuppression	Cas9 nuclease or Base editing	Improved persistence and tumor infiltration	Expanding KO library	[72–75]
Cytokine armoring	IL2, IL12 (CAR T), IL15 (CAR NK), HLA-E overexpression	Promoter-driven cytokine integration (retrovirus or KI)	Enhances persistence, combats TME suppression	Increases potency in solid tumors	[76–78]
iPSC-derived CAR NK	Dual anti-CD19/anti-BCMA CARs, KOs (B2M, CIITA, CD16), KI (HLA-E/G, IL2RF, tEGFR).	Multiplex Cas9 nuclease editing	Scalable, standardized CAR source; effective in systemic sclerosis model	Unlimited supply, safety switch built-in	[79]
HSPC-derived CAR NKT cells	Mesothelin CAR (MCAR) + IL15	Lentivirus	Stronger homing to ovarian cancer, reduced CRS toxicity	Outperformed CAR T-cells in solid tumors	[80]

Approach	Target or strategy	Editing tool	Outcome	Notes	Reference
Engineered bacterial support	E. coli expressing IL18 mutein (DR18)	Combined with mesothelin-CAR NK	Boosted survival and function in the TME; suppressed tumors in CRC and melanoma models	Bacteria-assisted cell therapy	[81]

Table 3.
Gene-editing strategies for cancer therapy.

mismatches between the sgRNA and the target DNA, potentially generating off-target editing in host cells. Such unintended effects may include indels, base edits, and even structural variations. Cas9 nucleases often produce more indels and chromosomal rearrangements due to DSBs. BE is efficient for point mutations but prone to bystander edits, while PE offers broader editing capacity with fewer off-target effects, though it is less efficient and may introduce errors via the RT. These safety issues have driven the development of various sensitive detection methods (in silico, in vitro, cell-based, in vivo) to better evaluate off-target effects [62, 63]. This is particularly important in contexts of multiplex gene editing, where multiple sgRNAs are used at the same time.

Continuous research refines CRISPR systems, enhances precision, minimizes off-target effects, and expands their applications in medicine, agriculture, and biotechnology. Enhancing the design of sgRNA, engineering high-fidelity Cas proteins, improving tissue-specific delivery, and limiting enzyme exposure are important strategies to reduce off-target effects and improve the safety of gene editing.

8. Application of gene editing in cancer

Gene editing offers powerful tools against cancer by precisely modifying genes in tumor cells or in immune cells that target cancer. CRISPR editing can inactivate oncogenes that stimulate tumor progression, such as *MYC*, or even reactivate tumor suppressors, like *TP53* that have been suppressed in cancer [64] (**Table 3**). Editing cancer driver mutations holds great promise for cancer therapy. However, current CRISPR-Cas9 tools often yield suboptimal editing rates, allowing residual unedited cancer cells to persist and revive the tumor.

An alternative CRISPR strategy to combat cancer lies in engineering autologous or allogeneic immune cells with a chimeric antigen receptor (CAR). CARs can recognize specific antigens, such as CD19 to eliminate B-cell malignancies [65], or BCMA to destroy multiple myeloma cells [66] (**Table 3**). CARs are typically inserted into T– or NK-cells due to their successful antigen-specific cytotoxicity against hematological malignancies. CAR T-cells usually exhibit prolonged expansion and persistence in patients, while CAR NK-cells present lower risks of graft-versus-host disease and cytokine release syndrome (CRS). The insertion of CARs into T–or NK-cell genomes can be achieved using nontargeted integration of retrovirus and lentivirus, or more recently, using safer Cas9-stimulated targeted knockin (KI) of rAAV templates into safe-harbor loci [67].

Because rAAV manufacturing is costly and time-consuming, with frequent observation of concatemeric insertions of rAAV vectors into the host genome [82], nonviral DNA represents an interesting alternative as a template for HDR-based KI of CARs. Naked dsDNA is typically more toxic to cells than rAAV or ssDNA at high concentrations [83]. However, a long ssDNA template flanked by short dsDNA Cas9 target sequences (ssCTS) has recently demonstrated reduced toxicity, potentially providing protection against DNA-sensing proteins [68] (**Table 3**). This ssCTS system showed editing efficiencies comparable to rAAV while offering simpler and more scalable production. Moreover, flanking dsDNA Cas9 sites with a partially mismatched sequence improved KI efficiency, suggesting that Cas9 can bind but not cleave the ssCTS and facilitate its entrance into the nucleus [68]. In this case, HiFi SpCas9 nuclease was used with an sgRNA for their targeted KI.

CAR T- and NK-cells are often engineered with the simultaneous knockout (KO) of multiple immune checkpoint genes to enhance their persistence and therapeutic efficacy. One of these genes is *TRAC*, whose KO prevents the recognition of recipient antigens by donor T-cell receptors (TCRs) and a graft-versus-host response. The insertion of a CD19-specific CAR into the *TRAC* locus allowed both the KO of *TRAC* and the expression of CD19 CAR under the transcriptional control of endogenous *TRAC*, which improved CAR T-cell persistence and potency [67] (**Table 3**).

To avoid simultaneous DSBs while silencing additional checkpoint genes and risking genomic rearrangements, SpCas9 nuclease was replaced with the ortholog AsCas12a Ultra (from *Acidaminococcus species*; with mutations that improve its performance [84]) for CAR KI [69] (**Table 3**). This strategy allowed the use of SpCas9-based ABE with additional sgRNAs to KO two more genes, that is, *B2M* and *CIITA*, preventing the recognition of MHC class I and II molecules by recipient immune cells and host-versus-graft response [69]. In this case, multiplex ABE was used to disrupt specific splice sites, leading to exon skipping and loss of functional domains. Since the sgRNAs used by SpCas9-ABE couldn't be recognized by the ortholog AsCas12a, a single DSB was created, thereby reducing the likelihood of genomic rearrangements.

A recent study has combined AsCas12a Ultra with an ssCTS template to improve the insertion of a few large transgenes into different loci in T-cells [85]. This approach led to an impressive 90% insertion efficiency of HLA-A2 TRuC into *CD3E*, easily matching rAAV in efficiency. The authors also reported a 35% insertion efficiency of CD19 CAR into *TRAC*. These results demonstrate robust and precise T-cell gene editing using virus-free conditions involving AsCas12a Ultra and large ssCTS templates. It remains to be shown how this strategy is compatible with simultaneous multiplex KO using SpCas9-ABEs.

A growing list of gene KOs has been shown to enhance T-cell persistence and proliferative capacity, in addition to *TRAC*, *B2M* and *CIITA*. For example, *PDCD1* is involved in inhibitory immune checkpoints [86], *CD3E* is involved in TCR complex expression [70], and *PVR* may mitigate the missing-self issue [70]. ABE8e mRNA demonstrated higher efficiency and fewer translocations than HiFi SpCas9 nuclease (due to fewer DSBs) during the simultaneous quadruplex KO of *CD3E*, *B2M*, *CIITA*, and *PVR* in CAR T-cells [70] (**Table 3**). With editing efficiencies ranging from 85% to 95% for all these four targets, the generated CAR T-cells showed boosted persistence and antitumor activity in vivo [70].

Similarly, multiplex KO using BE can significantly enhance CAR NK-cell function. For example, ABE8e mRNA was used to simultaneously destroy splice donors in *AHR*, *CISH*, *KLRG1*, *TIGIT*, and *PDCD1*, showing 65–100% editing efficiencies, undetectable chromosomal translocations, and improved function of CAR NK-cells [71] (**Table 3**). Disrupting *CD38* expression was shown to prevent T- and NK-cell fratricide, suggesting that this gene may be a valuable addition to the multiplex KO design [72, 73] (**Table 3**). Likewise, in CAR NK-cells, the KO of *CALHM2* enhances cytotoxicity and tumor infiltration [74], and the KO of *CREM* boosts their resistance to tumor-induced immunosuppression [75] (**Table 3**).

The efficacy of CAR T- and NK-cells can also be improved by armoring them with cytokines to combat the immunosuppressive environment in solid tumors. For example, *IL2* or *IL12* expression, driven by the promoters of *NR4A2* or *RGS16* in CAR T-cells, increases their antitumor capacity and persistence [76] (**Table 3**). In CAR NK-cells, similar improvements were observed with *IL15* overexpressed using retrovirus [78] (**Table 3**). Additionally, *HLA-E* overexpression in NK-cells was shown to abrogate alloreactive NK-cell clearance [77] (**Table 3**).

CAR T- and NK-cells can also be differentiated from human induced pluripotent stem cells (iPSCs), which are amenable to clonal multiplex gene editing and offer a limitless cell source. For example, engineering iPSCs at multiple genetic loci can produce safe and efficient CAR NK-cells for the treatment of severe refractory systemic sclerosis [79]. This strategy involved the dual insertion of anti-CD19 and anti-BCMA CARs, the KO of *B2M*, *CIITA,* and *CD16* to reduce rejection by host immune cells, in addition to the KI of *HLA-E*, *HLA-G*, *IL-2RF* to enhance survival, and *tEGFR* to provide a safety switch for transplanted cells [79] (**Table 3**). The ssCTS template strategy presented above, combined with Cas9, should be considered safer than viruses for use in the future to equip CAR cells with different armors.

Furthermore, allogeneic CAR NKT-cells derived from multiplex gene-edited hematopoietic stem and progenitor cells (HSPCs) have demonstrated encouraging results in the fight against solid tumors like ovarian cancer (OC) [80]. These cells were engineered using lentivirus to target mesothelin (MCAR) and express *IL15* (**Table 3**). They could outperform CAR T-cells in OC tumor homing and counteract the immune-suppressive tumor microenvironment (TME), with reduced CRS toxicity [80].

Interestingly, some bacteria may help CAR NK-cells resist the TME surrounding solid tumors. For example, a nonpathogenic *E. coli* strain was engineered to express a decoy-resistant IL18 mutein (DR18) on their surface, helping mesothelin-targeting CAR NK-cells survive in the TME [81] (**Table 3**). This approach led to the suppression of tumor progression in immune-competent colorectal carcinoma and melanoma mouse models.

Next-generation cell and gene therapies to fight cancer are poised to incorporate a wide array of customized gene edits to optimize their efficacy. This is critical in the context of solid tumors, which present arduous challenges due to their immunosuppressive microenvironment, physical barriers, and tumor heterogeneity. Achieving durable and potent responses will likely depend on multiplex gene editing strategies that enhance immune cell persistence and tumor infiltration, resist exhaustion, and overcome local immunosuppression.

9. Cure of genetic diseases beyond cancer

Gene editing also offers a direct means to correct genetic mutations underlying numerous inherited diseases, as extensively documented [87]. The ultimate goal of gene therapies is to fix the root cause of genetic disorders through single-dose cures with life-lasting effects. The world's first approved CRISPR-based gene editing therapy, by the UK Medicines and Healthcare Products Regulatory Agency, was Casgevy™ (exagamglogene autotemcel) in 2023. It treats SCD and transfusion-dependent beta-thalassemia (TDT) by editing a DNA sequence in patients' blood stem cells ex vivo to restore fetal hemoglobin (HbF) production [4, 88]. This therapy uses SpCas9 nuclease and a specific sgRNA to disrupt a *BCL11A* enhancer via NHEJ-derived indels, derepressing gamma-globin and HbF expression. Clinical trials showed ~ 80% editing efficiency and significant HbF induction without detectable off-target effects (e.g., NCT03655678, NCT03745287 from clinicaltrials.gov).

BE was also leveraged, using the highly active variant of ABE8e, to correct a pathogenic SCD mutation in the beta-globin gene (*HBB*) in patient-derived $CD34^+$ cells [89]. Upon transplantation into humanized mice, the editing achieved 68% efficiency, with minimal bystander edits and only a few off-target mutations, none of which were expected to have clinical relevance [89].

Furthermore, another group achieved in vivo PE of *HBB* using the PE5max system in an SCD mouse model following HSC mobilization [90]. This approach resulted in 40% editing efficiency, 43% hemoglobin replacement, reduced SCD-related symptoms, 1.5% on-target indels, and no detectable off-target effects [90].

A growing number of gene editing tools are being applied in both preclinical and clinical trials to target a broad spectrum of previously incurable genetic disorders. From monogenic to complex polygenic diseases, therapeutic gene editing is progressing rapidly and yielding promising results. Notably, the correction of pathogenic mutations is addressing multiple conditions, including SCID [91], chronic granulomatous disease [92], hemophilia [93], Alzheimer's disease [94], Huntington's disease [95], Niemann-Pick disease type C [96], alternating hemiplegia of childhood [97], ophthalmic disorders (NCT03872479) [98], congenital hearing loss (NCT06025032) [99], Duchenne muscular dystrophy [100], spinal muscular atrophy [101], cardiomyopathy [102], Hutchinson-Gilford progeria syndrome [103], hypercholesterolemia (NCT05398029) [104], transthyretin amyloidosis (NCT04601051) [105], tyrosinemia [106], alpha-1 antitrypsin deficiency [107], phenylketonuria [108], primary hyperoxaluria type 1 [109], and hereditary angioedema [110].

As gene-editing technologies continue to progress, a growing number of genetic diseases are expected to become curable in the near future, ultimately reducing lifelong suffering, dependency on chronic medication, and the burden on public health systems.

10. Ethics and regulation

Gene editing holds great promise in medicine but also raises ethical concerns when we envision its immense potential. The responsible use and potential misuse of gene editing technologies remain matters of intense debate in the scientific community. Most of the therapeutic gene editing presented above is, hopefully, limited to somatic (nonreproductive) cells and cannot be transmitted to the next generation. However, germline editing is heritable and may lead to unexpected genetic issues, such as inherited off-target effects. This may lead to ethical questions about the lack of consent from the descendants who will inherit those edits. For example, the controversial revelation of the existence of the "CRISPR babies" in 2018, whose genomes had been edited at the early embryo stage to KO *CCR5* and confer innate resistance to human immunodeficiency virus (HIV) infection [111], has accelerated international discussions about ethical issues associated with heritable human genome editing and enhancement. The probable long-term consequences of germline transmission face strong resistance from scientists, ethicists, and the public. Germline editing is actually prohibited or heavily restricted in several countries around the world [112].

Regulatory agencies worldwide are crucial for gene editing oversight and to ensure its safe application in human health. For example, an LNP-based CRISPR delivery system that targets somatic cells but exhibits such low specificity that it may also edit germ cells warrants rigorous investigation. In this case, improving nanoparticle biocompatibility is necessary to reduce unexpected damage. Regulatory frameworks are pivotal in balancing gene editing innovation with safety concerns.

11. Conclusion

The history of gene editing follows a steady progression from the discovery of DNA to the engineering of precise and versatile tools. Pioneer-engineered nucleases made targeted gene manipulation a reality, but their complexity restricted widespread development. The discovery of the CRISPR system, particularly Cas9, revolutionized the field by providing simplicity, programmability, and broad applicability. Further evolution has produced more specialized variants of Cas enzymes with improved specificity and greater editing capacities, such as base and PE. These advances have enabled spectacular applications across medicine, agriculture, and biotechnology. In cancer, CRISPR has opened the way to targeted oncogene disruption and powerful immune cell engineering. In genetic diseases, it has led to new curative strategies like Casgevy™ for SCD and beta-thalassemia. Despite safety obstacles such as off-target effects and delivery limitations, constant innovations will hopefully soon provide safe, efficient, and more accessible therapies. Overall, gene editing has progressed from a conceptual dream to a possible and rapidly expanding therapeutic reality with profound implications for human health.

Acknowledgments

This work was supported by the Government of Canada.

Conflict of interest

The authors declare no conflict of interest.

DOI: http://dx.doi.org/10.5772/intechopen.1013143

Author details

Eric Deneault
Regulatory Research Division, Centre for Oncology, Radiopharmaceuticals and Research, Biologic and Radiopharmaceutical Drugs Directorate, Health Products and Food Branch, Health Canada, Ottawa, Canada

*Address all correspondence to: eric.deneault@hc-sc.gc.ca

References

[1] Dahm R. Discovering DNA: Friedrich Miescher and the early years of nucleic acid research. Human Genetics. 2008;**122**:565–581. DOI: 10.1007/s00439-007-0433-0

[2] Avery OT, Macleod CM, McCarty M. Studies on the Chemical Nature of the Substance Inducing Transformation of Pneumococcal Types: Induction of Transformation by a Desoxyribonucleic Acid Fraction Isolated from Pneumococcus Type III. Journal of Experimental Medicine. 1944;**79**:137–158. DOI: 10.1084/jem.79.2.137

[3] Watson JD, Crick FH. Molecular structure of nucleic acids; a structure for deoxyribose nucleic acid. Nature. 1953;**171**:737–738. DOI: 10.1038/171737a0

[4] Sheridan C. The world's first CRISPR therapy is approved: Who will receive it? Nature Biotechnology. 2024;**42**:3–4. DOI: 10.1038/d41587-023-00016-6

[5] Smith HO, Wilcox KW. A restriction enzyme from Hemophilus influenzae. I. Purification and general properties. Journal of Molecular Biology. 1970;**51**:379–391. DOI: 10.1016/0022-2836(70)90149-x

[6] Kim YG, Cha J, Chandrasegaran S. Hybrid restriction enzymes: Zinc finger fusions to Fok I cleavage domain. Proceedings of the National Academy of Sciences of the United States of America. 1996;**93**:1156–1160. DOI: 10.1073/pnas.93.3.1156

[7] Urnov FD, Miller JC, Lee YL, Beausejour CM, Rock JM, Augustus S, Jamieson AC, Porteus MH, Gregory PD, Holmes MC. Highly efficient endogenous human gene correction using designed zinc-finger nucleases. Nature. 2005;**435**:646–651. DOI: 10.1038/nature03556

[8] Boch J, Scholze H, Schornack S, Landgraf A, Hahn S, Kay S, Lahaye T, Nickstadt A, Bonas U. Breaking the code of DNA binding specificity of TAL-type III effectors. Science. 2009;**326**:1509–1512. DOI: 10.1126/science.1178811

[9] Moscou MJ, Bogdanove AJ. A simple cipher governs DNA recognition by TAL effectors. Science. 2009;**326**:1501. DOI: 10.1126/science.1178817

[10] Ishino Y, Shinagawa H, Makino K, Amemura M, Nakata A. Nucleotide sequence of the iap gene, responsible for alkaline phosphatase isozyme conversion in Escherichia coli, and identification of the gene product. Journal of Bacteriology. 1987;**169**: 5429–5433. DOI: 10.1128/jb.169.12.5429-5433.1987

[11] Jansen R, Embden JD, Gaastra W, Schouls LM. Identification of genes that are associated with DNA repeats in prokaryotes. Molecular Microbiology. 2002;**43**:1565–1575. DOI: 10.1046/j.1365-2958.2002.02839.x

[12] Mojica FJ, Diez-Villasenor C, Garcia-Martinez J, Soria E. Intervening sequences of regularly spaced prokaryotic repeats derive from foreign genetic elements. Journal of Molecular Evolution. 2005;**60**:174–182. DOI: 10.1007/s00239-004-0046-3

[13] Jinek M, Chylinski K, Fonfara I, Hauer M, Doudna JA, Charpentier E.

A programmable dual-RNA-guided DNA endonuclease in adaptive bacterial immunity. Science. 2012;**337**:816–821. DOI: 10.1126/science.1225829

[14] Nishimasu H, Ran FA, Hsu PD, Konermann S, Shehata SI, Dohmae N, Ishitani R, Zhang F, Nureki O. Crystal structure of Cas9 in complex with guide RNA and target DNA. Cell. 2014;**156**:935–949. DOI: 10.1016/j.cell.2014.02.001

[15] Cong L, Ran FA, Cox D, Lin S, Barretto R, Habib N, Hsu PD, Wu X, Jiang W, Marraffini LA, Zhang F. Multiplex genome engineering using CRISPR/Cas systems. Science. 2013;**339**:819–823. DOI: 10.1126/science.1231143

[16] Jinek M, East A, Cheng A, Lin S, Ma E, Doudna J. RNA-programmed genome editing in human cells. Elife. 2013;**2**:e00471. DOI: 10.7554/eLife.00471

[17] Mali P, Yang L, Esvelt KM, Aach J, Guell M, DiCarlo JE, Norville JE, Church GM. RNA-guided human genome engineering via Cas9. Science. 2013;**339**:823–826. DOI: 10.1126/science.1232033

[18] Gasiunas G, Young JK, Karvelis T, Kazlauskas D, Urbaitis T, Jasnauskaite M, Grusyte MM, Paulraj S, Wang PH, Hou Z, et al. A catalogue of biochemically diverse CRISPR-Cas9 orthologs. Nature Communications. 2020;**11**:5512

[19] Hou Z, Zhang Y, Propson NE, Howden SE, Chu LF, Sontheimer EJ, Thomson JA. Efficient genome engineering in human pluripotent stem cells using Cas9 from Neisseria meningitidis. Proceedings of the National Academy of Sciences of the United States of America. 2013;**110**:15644–15649. DOI: 10.1073/pnas.1313587110

[20] Ran FA, Cong L, Yan WX, Scott DA, Gootenberg JS, Kriz AJ, Zetsche B, Shalem O, Wu X, Makarova KS, et al. In vivo genome editing using Staphylococcus aureus Cas9. Nature. 2015;**520**:186–191

[21] Kim E, Koo T, Park SW, Kim D, Kim K, Cho HY, Song DW, Lee KJ, Jung MH, Kim S, et al. In vivo genome editing with a small Cas9 orthologue derived from Campylobacter jejuni. Nature Communications. 2017;**8**:14500. DOI: 10.1038/ncomms14500

[22] Harrington LB, Paez-Espino D, Staahl BT, Chen JS, Ma E, Kyrpides NC, Doudna JA. A thermostable Cas9 with increased lifetime in human plasma. Nature Communications. 2017;**8**:1424. DOI: 10.1038/s41467-017-01408-4

[23] Deneault E, Faheem M, White SH, Rodrigues DC, Sun S, Wei W, Piekna A, Thompson T, Howe JL, Chalil L, et al. CNTN5(-)(/+)or EHMT2(-)(/+)human iPSC-derived neurons from individuals with autism develop hyperactive neuronal networks. Elife. 2019;**8**. DOI: 10.7554/eLife.40092

[24] Deneault E, White SH, Rodrigues DC, Ross PJ, Faheem M, Zaslavsky K, Wang Z, Alexandrova R, Pellecchia G, Wei W, et al. Complete disruption of autism-susceptibility genes by gene editing predominantly reduces functional connectivity of isogenic human neurons. Stem Cell Reports. 2018;**11**:1211–1225

[25] Murtaza N, Cheng AA, Brown CO, Meka DP, Hong S, Uy JA, El-Hajjar J, Pipko N, Unda BK, Schwanke B, et al. Neuron-specific protein network

mapping of autism risk genes identifies shared biological mechanisms and disease-relevant pathologies. Cell Reports. 2022;**41**:111678. DOI: 10.1016/j.celrep.2022.111678

[26] Ran FA, Hsu PD, Lin CY, Gootenberg JS, Konermann S, Trevino AE, Scott DA, Inoue A, Matoba S, Zhang Y, Zhang F. Double nicking by RNA-guided CRISPR Cas9 for enhanced genome editing specificity. Cell. 2013;**154**:1380–1389. DOI: 10.1016/j.cell.2013.08.021

[27] Ross PJ, Zhang WB, Mok RSF, Zaslavsky K, Deneault E, D'Abate L, Rodrigues DC, Yuen RKC, Faheem M, Mufteev M, et al. Synaptic Dysfunction in Human Neurons With Autism-Associated Deletions in PTCHD1-AS. Biological Psychiatry. 2020;**87**:139–149

[28] Woodbury-Smith M, Deneault E, Yuen RKC, Walker S, Zarrei M, Pellecchia G, Howe JL, Hoang N, Uddin M, Marshall CR, et al. Mutations in RAB39B in individuals with intellectual disability, autism spectrum disorder, and macrocephaly. Molecular Autism. 2017;**8**:59

[29] Zaslavsky K, Zhang WB, McCready FP, Rodrigues DC, Deneault E, Loo C, Zhao M, Ross PJ, El Hajjar J, Romm A, et al. SHANK2 mutations associated with autism spectrum disorder cause hyperconnectivity of human neurons. Nature Neuroscience. 2019;**22**:556–564

[30] Komor AC, Kim YB, Packer MS, Zuris JA, Liu DR. Programmable editing of a target base in genomic DNA without double-stranded DNA cleavage. Nature. 2016;**533**:420–424. DOI: 10.1038/nature17946

[31] Gaudelli NM, Komor AC, Rees HA, Packer MS, Badran AH, Bryson DI, Liu DR. Programmable base editing of A*T to G*C in genomic DNA without DNA cleavage. Nature. 2017;**551**:464–471. DOI: 10.1038/nature24644

[32] Grunewald J, Zhou R, Garcia SP, Iyer S, Lareau CA, Aryee MJ, Joung JK. Transcriptome-wide off-target RNA editing induced by CRISPR-guided DNA base editors. Nature. 2019;**569**:433–437. DOI: 10.1038/s41586-019-1161-z

[33] Zhou C, Sun Y, Yan R, Liu Y, Zuo E, Gu C, Han L, Wei Y, Hu X, Zeng R, et al. Off-target RNA mutation induced by DNA base editing and its elimination by mutagenesis. Nature. 2019;**571**:275–278

[34] Richter MF, Zhao KT, Eton E, Lapinaite A, Newby GA, Thuronyi BW, Wilson C, Koblan LW, Zeng J, Bauer DE, et al. Phage-assisted evolution of an adenine base editor with improved Cas domain compatibility and activity. Nature Biotechnology. 2020;**38**:883–891

[35] Anzalone AV, Randolph PB, Davis JR, Sousa AA, Koblan LW, Levy JM, Chen PJ, Wilson C, Newby GA, Raguram A, Liu DR. Search-and-replace genome editing without double-strand breaks or donor DNA. Nature. 2019;**576**:149–157. DOI: 10.1038/s41586-019-1711-4

[36] Chen PJ, Hussmann JA, Yan J, Knipping F, Ravisankar P, Chen PF, Chen C, Nelson JW, Newby GA, Sahin M, et al. Enhanced prime editing systems by manipulating cellular determinants of editing outcomes. Cell. 2021;**184**:5635–5652e5629

[37] Doman JL, Pandey S, Neugebauer ME, An M, Davis JR, Randolph PB, McElroy A, Gao XD, Raguram A, Richter MF, et al. Phage-assisted evolution and protein engineering yield compact, efficient prime editors. Cell. 2023;**186**: 3983–4002e3926

[38] Qi LS, Larson MH, Gilbert LA, Doudna JA, Weissman JS, Arkin AP, Lim WA. Repurposing CRISPR as an RNA-guided platform for sequence-specific control of gene expression. Cell. 2013;**152**:1173–1183. DOI: 10.1016/j.cell.2013.02.022

[39] Gilbert LA, Larson MH, Morsut L, Liu Z, Brar GA, Torres SE, Stern-Ginossar N, Brandman O, Whitehead EH, Doudna JA, et al. CRISPR-mediated modular RNA-guided regulation of transcription in eukaryotes. Cell. 2013;**154**:442–451

[40] Kleinstiver BP, Pattanayak V, Prew MS, Tsai SQ, Nguyen NT, Zheng Z, Joung JK. High-fidelity CRISPR-Cas9 nucleases with no detectable genome-wide off-target effects. Nature. 2016;**529**:490–495. DOI: 10.1038/nature16526

[41] Vakulskas CA, Dever DP, Rettig GR, Turk R, Jacobi AM, Collingwood MA, Bode NM, McNeill MS, Yan S, Camarena J, et al. A high-fidelity Cas9 mutant delivered as a ribonucleoprotein complex enables efficient gene editing in human hematopoietic stem and progenitor cells. Nature Medicine. 2018;**24**:1216–1224

[42] Slaymaker IM, Gao L, Zetsche B, Scott DA, Yan WX, Zhang F. Rationally engineered Cas9 nucleases with improved specificity. Science. 2016;**351**:84–88. DOI: 10.1126/science.aad5227

[43] Chen JS, Dagdas YS, Kleinstiver BP, Welch MM, Sousa AA, Harrington LB, Sternberg SH, Joung JK, Yildiz A, Doudna JA. Enhanced proofreading governs CRISPR-Cas9 targeting accuracy. Nature. 2017;**550**:407–410. DOI: 10.1038/nature24268

[44] Hu JH, Miller SM, Geurts MH, Tang W, Chen L, Sun N, Zeina CM, Gao X, Rees HA, Lin Z, Liu DR. Evolved Cas9 variants with broad PAM compatibility and high DNA specificity. Nature. 2018;**556**:57–63. DOI: 10.1038/nature26155

[45] Nishimasu H, Shi X, Ishiguro S, Gao L, Hirano S, Okazaki S, Noda T, Abudayyeh OO, Gootenberg JS, Mori H, et al. Engineered CRISPR-Cas9 nuclease with expanded targeting space. Science. 2018;**361**:1259–1262

[46] Walton RT, Christie KA, Whittaker MN, Kleinstiver BP. Unconstrained genome targeting with near-PAMless engineered CRISPR-Cas9 variants. Science. 2020;**368**:290–296. DOI: 10.1126/science.aba8853

[47] Yan J, Oyler-Castrillo P, Ravisankar P, Ward CC, Levesque S, Jing Y, Simpson D, Zhao A, Li H, Yan W, et al. Improving prime editing with an endogenous small RNA-binding protein. Nature. 2024;**628**:639–647

[48] Zetsche B, Gootenberg JS, Abudayyeh OO, Slaymaker IM, Makarova KS, Essletzbichler P, Volz SE, Joung J, van der Oost J, Regev A, et al. Cpf1 is a single RNA-guided endonuclease of a class 2 CRISPR-Cas system. Cell. 2015;**163**:759–771

[49] Chen JS, Ma E, Harrington LB, Da Costa M, Tian X, Palefsky JM,

Doudna JA. CRISPR-Cas12a target binding unleashes indiscriminate single-stranded DNase activity. Science. 2018;**360**:436–439. DOI: 10.1126/science.aar6245

[50] Abudayyeh OO, Gootenberg JS, Konermann S, Joung J, Slaymaker IM, Cox DB, Shmakov S, Makarova KS, Semenova E, Minakhin L, et al. C2c2 is a single-component programmable RNA-guided RNA-targeting CRISPR effector. Science. 2016;**353**:aaf5573

[51] Villiger L, Joung J, Koblan L, Weissman J, Abudayyeh OO, Gootenberg JS. CRISPR technologies for genome, epigenome and transcriptome editing. Nature Reviews Molecular Cell Biology. 2024. DOI: 10.1038/s41580-023-00697-6

[52] Gootenberg JS, Abudayyeh OO, Lee JW, Essletzbichler P, Dy AJ, Joung J, Verdine V, Donghia N, Daringer NM, Freije CA, et al. Nucleic acid detection with CRISPR-Cas13a/C2c2. Science. 2017;**356**:438–442

[53] Harrington LB, Burstein D, Chen JS, Paez-Espino D, Ma E, Witte IP, Cofsky JC, Kyrpides NC, Banfield JF, Doudna JA. Programmed DNA destruction by miniature CRISPR-Cas14 enzymes. Science. 2018;**362**:839–842. DOI: 10.1126/science.aav4294

[54] Frischmeyer PA, Dietz HC. Nonsense-mediated mRNA decay in health and disease . Human Molecular Genetics. 1999;**8**:1893–1900. DOI: 10.1093/hmg/8.10.1893

[55] Chen CX, Deneault E, Abdian N, You Z, Sirois J, Nicouleau M, Shlaifer I, Villegas L, Boivin MN, Gaborieau L, et al. Generation of patient-derived pluripotent stem cell-lines and CRISPR modified isogenic controls with mutations in the Parkinson's associated GBA gene. Stem Cell Research. 2022;**64**:102919. DOI: 10.1016/j.scr.2022.102919

[56] Deneault E, Chaineau M, Nicouleau M, Castellanos Montiel MJ, Franco Flores AK, Haghi G, Chen CX, Abdian N, Shlaifer I, Beitel LK, Durcan TM. A streamlined CRISPR workflow to introduce mutations and generate isogenic iPSCs for modeling amyotrophic lateral sclerosis. Methods. 2022;**203**:297–310. DOI: 10.1016/j.ymeth.2021.09.002

[57] Faheem M, Deneault E, Alexandrova R, Rodrigues DC, Pellecchia G, Shum C, Zarrei M, Piekna A, Wei W, Howe JL, et al. Disruption of DDX53 coding sequence has limited impact on iPSC-derived human NGN2 neurons. BMC Medical Genomics. 2023;**16**:5

[58] Fiesel FC, Bustillos BA, Watzlawik JO, Chen CXQ, Berryer MH, Zhang J, Boneski PK, Hayes CS, Bredenberg JM, Deneault E, et al. Activation of endogenous PRKN by structural derepression is linked to increased turnover of the E3 ubiquitin ligase. Autophagy. 2025:1–21. DOI: 10.1080/15548627.2025.2531025

[59] Lepine S, Nauleau-Javaudin A, Deneault E, Chen CX, Abdian N, Franco-Flores AK, Haghi G, Castellanos-Montiel MJ, Maussion G, Chaineau M, Durcan TM. Homozygous ALS-linked mutations in TARDBP/TDP-43 lead to hypoactivity and synaptic abnormalities in human iPSC-derived motor neurons. iScience. 2024;**27**:109166. DOI: 10.1016/j.isci.2024.109166

[60] Maussion G, Rocha C, Abdian N, Yang D, Turk J, Carrillo

Valenzuela D, Pimentel L, You Z, Morquette B, Nicouleau M, et al. Transcriptional Dysregulation and Impaired Neuronal Activity in FMR1 Knock-Out and Fragile X Patients' iPSC-Derived Models. International Journal of Molecular Sciences. 2023;**24**:14926

[61] Mohamed NV, Sirois J, Ramamurthy J, Mathur M, Lepine P, Deneault E, Maussion G, Nicouleau M, Chen CX, Abdian N, et al. Midbrain organoids with an SNCA gene triplication model key features of synucleinopathy. Brain Communications. 2021;**3**:fcab223

[62] Guo C, Ma X, Gao F, Guo Y. Off-target effects in CRISPR/Cas9 gene editing. Frontiers in Bioengineering and Biotechnology. 2023;**11**:1143157. DOI: 10.3389/fbioe.2023.1143157

[63] Hunt JMT, Samson CA, Rand AD, Sheppard HM. Unintended CRISPR-Cas9 editing outcomes: A review of the detection and prevalence of structural variants generated by gene-editing in human cells. Human Genetics. 2023;**142**: 705–720. DOI: 10.1007/s00439-023-02561-1

[64] Chehelgerdi M, Chehelgerdi M, Khorramian-Ghahfarokhi M, Shafieizadeh M, Mahmoudi E, Eskandari F, Rashidi M, Arshi A, Mokhtari-Farsani A. Comprehensive review of CRISPR-based gene editing: Mechanisms, challenges, and applications in cancer therapy. Molecular Cancer. 2024;**23**:9. DOI: 10.1186/s12943-023-01925-5

[65] Sadelain M. CAR therapy: The CD19 paradigm. Journal of Clinical Investigation. 2015;**125**:3392–3400. DOI: 10.1172/JCI80010

[66] Ershova A, Goldaeva A, Staliarova A, Bulatov E, Petukhov A, Barlev N. Future perspectives on novel CAR-T therapeutics beyond CD19 and BCMA in onco-hematology. Frontiers in Immunology. 2025;**16**:1592377. DOI: 10.3389/fimmu.2025.1592377

[67] Eyquem J, Mansilla-Soto J, Giavridis T, van der Stegen SJ, Hamieh M, Cunanan KM, Odak A, Gonen M, Sadelain M. Targeting a CAR to the TRAC locus with CRISPR/Cas9 enhances tumour rejection. Nature. 2017;**543**: 113–117. DOI: 10.1038/nature21405

[68] Shy BR, Vykunta VS, Ha A, Talbot A, Roth TL, Nguyen DN, Pfeifer WG, Chen YY, Blaeschke F, Shifrut E, et al. High-yield genome engineering in primary cells using a hybrid ssDNA repair template and small-molecule cocktails. Nature Biotechnology. 2023;**41**:521–531

[69] Glaser V, Flugel C, Kath J, Du W, Drosdek V, Franke C, Stein M, Pruss A, Schmueck-Henneresse M, Volk HD, et al. Combining different CRISPR nucleases for simultaneous knock-in and base editing prevents translocations in multiplex-edited CAR T cells. Genome Biology. 2023;**24**:89

[70] Engel NW, Steinfeld I, Ryan D, Anupindi K, Kim S, Wellhausen N, Chen L, Wilkins K, Baker DJ, Rommel PC, et al. Quadruple adenine base-edited allogeneic CAR T cells outperform CRISPR/Cas9 nuclease-engineered T cells. Proceedings of the National Academy of Sciences of the United States of America. 2025;**122**: e2427216122

[71] Wang M, Krueger JB, Gilkey AK, Stelljes EM, Kluesner MG, Pomeroy EJ, Skeate JG, Slipek NJ, Lahr WS, Claudio Vazquez PN, et al. Precision

enhancement of CAR-NK cells through non-viral engineering and highly multiplexed base editing. Journal for ImmunoTherapy of Cancer. 2025;**13**: e009560

[72] Gurney M, Stikvoort A, Nolan E, Kirkham-mccarthy L, Khoruzhenko S, Shivakumar R, Zweegman S, Van de Donk N, Mutis T, Szegezdi E, et al. CD38 knockout natural killer cells expressing an affinity optimized CD38 chimeric antigen receptor successfully target acute myeloid leukemia with reduced effector cell fratricide. Haematologica. 2022;**107**:437–445

[73] Liao C, Wang Y, Huang Y, Duan Y, Liang Y, Chen J, Jiang J, Shang K, Zhou C, Gu Y, et al. CD38-Specific CAR Integrated into CD38 Locus Driven by Different Promoters Causes Distinct Antitumor Activities of T and NK Cells. Advanced Science. 2023;**10**:e2207394

[74] Peng L, Renauer PA, Sferruzza G, Yang L, Zou Y, Fang Z, Park JJ, Chow RD, Zhang Y, Lin Q, et al. In vivo AAV-SB-CRISPR screens of tumor-infiltrating primary NK cells identify genetic checkpoints of CAR-NK therapy. Nature Biotechnology. 2025;**43**:752–761

[75] Rafei H, Basar R, Acharya S, Hsu YS, Liu P, Zhang D, Bohn T, Liang Q, Mohanty V, Upadhyay R, et al. CREM is a regulatory checkpoint of CAR and IL-15 signalling in NK cells. Nature. 2025;**643**:1076–1086

[76] Chen AXY, Yap KM, Kim JS, Sek K, Huang YK, Dunbar PA, Wiebking V, Armitage JD, Munoz I, Todd KL, et al. Rewiring endogenous genes in CAR T cells for tumour-restricted payload delivery. Nature. 2025;**644**:241–251

[77] Huang RS, Lai MC, Shih HA, Lin S. A robust platform for expansion and genome editing of primary human natural killer cells. Journal of Experimental Medicine. 2021;**218**. DOI: 10.1084/jem.20201529

[78] Liu E, Tong Y, Dotti G, Shaim H, Savoldo B, Mukherjee M, Orange J, Wan X, Lu X, Reynolds A, et al. Cord blood NK cells engineered to express IL-15 and a CD19-targeted CAR show long-term persistence and potent antitumor activity. Leukemia. 2018;**32**:520–531

[79] Wang X, Zhang Y, Jin Y, Dai L, Yue Y, Hu J, Liu X, Pang K, Ye S, Chen Y, et al. An iPSC-derived CD19/BCMA CAR-NK therapy in a patient with systemic sclerosis. Cell. 2025;**188**:4225–4238e4212

[80] Li YR, Li Z, Zhu Y, Li M, Chen Y, Lee D, Ochoa CJ, Singh T, DiBernardo G, Guo W, et al. Overcoming ovarian cancer resistance and evasion to CAR-T cell therapy by harnessing allogeneic CAR-NKT cells. Med. 2025;100804. DOI: 10.1016/j.medj.2025.100804

[81] Yang S, Sheffer M, Kaplan IE, Wang Z, Tarannum M, Dinh K, Abdulhamid Y, Bobilev E, Shapiro R, Porter R, et al. Non-pathogenic E. coli displaying decoy-resistant IL18 mutein boosts anti-tumor and CAR NK cell responses. Nature Biotechnology. 2025;**43**:1311–1323

[82] Suchy FP, Karigane D, Nakauchi Y, Higuchi M, Zhang J, Pekrun K, Hsu I, Fan AC, Nishimura T, Charlesworth CT, et al. Genome engineering with Cas9 and AAV repair templates generates frequent concatemeric insertions of viral vectors. Nature Biotechnology. 2025;**43**:204–213

[83] Roth TL, Puig-Saus C, Yu R, Shifrut E, Carnevale J, Li PJ, Hiatt J, Saco J, Krystofinski P, Li H, et al. Reprogramming human T cell function and specificity with non-viral genome targeting. Nature. 2018;**559**:405–409

[84] Zhang L, Zuris JA, Viswanathan R, Edelstein JN, Turk R, Thommandru B, Rube HT, Glenn SE, Collingwood MA, Bode NM, et al. AsCas12a ultra nuclease facilitates the rapid generation of therapeutic cell medicines. Nature Communications. 2021;**12**:3908

[85] Nitulescu AM, Du W, Glaser V, Kath J, Aird EJ, Cullot G, Greensmith R, Mikkelsen NS, Stein M, Bak RO, et al. Single-stranded HDR templates with truncated Cas12a-binding sequences improve knock-in efficiencies in primary human T cells. Molecular Therapy Nucleic Acids. 2025;**36**:102568. DOI: 10.1016/j.omtn.2025.102568

[86] Menger L, Sledzinska A, Bergerhoff K, Vargas FA, Smith J, Poirot L, Pule M, Hererro J, Peggs KS, Quezada SA. TALEN-Mediated Inactivation of PD-1 in Tumor-Reactive Lymphocytes Promotes Intratumoral T-cell Persistence and Rejection of Established Tumors. Cancer Research. 2016;**76**:2087–2093. DOI: 10.1158/0008-5472.CAN-15-3352

[87] Deneault E. Recent Therapeutic Gene Editing Applications to Genetic Disorders. Current Issues in Molecular Biology. 2024;**46**:4147–4185. DOI: 10.3390/cimb46050255

[88] Frangoul H, Altshuler D, Cappellini MD, Chen YS, Domm J, Eustace BK, Foell J, de la Fuente J, Grupp S, Handgretinger R, et al. CRISPR-Cas9 Gene Editing for Sickle Cell Disease and beta-Thalassemia. New England Journal of Medicine. 2021;**384**:252–260

[89] Newby GA, Yen JS, Woodard KJ, Mayuranathan T, Lazzarotto CR, Li Y, Sheppard-Tillman H, Porter SN, Yao Y, Mayberry K, et al. Base editing of haematopoietic stem cells rescues sickle cell disease in mice. Nature. 2021;**595**:295–302

[90] Li C, Georgakopoulou A, Newby GA, Chen PJ, Everette KA, Paschoudi K, Vlachaki E, Gil S, Anderson AK, Koob T, et al. In vivo HSC prime editing rescues sickle cell disease in a mouse model. Blood. 2023;**141**:2085–2099

[91] McAuley GE, Yiu G, Chang PC, Newby GA, Campo-Fernandez B, Fitz-Gibbon ST, Wu X, Kang SL, Garibay A, Butler J, et al. Human T cell generation is restored in CD3delta severe combined immunodeficiency through adenine base editing. Cell. 2023;**186**: 1398–1416e1323

[92] Heath JM, Orenstein JS, Tedeschi JG, Ng A, Collier MD, Kushakji J, Wilhelm AJ, Taylor A, DP W, De Ravin SS, et al. Prime Editing Efficiently and Precisely Corrects Causative Mutation in Chronic Granulomatous Disease, Restoring Myeloid Function: Toward Development of a Prime Edited Autologous Hematopoietic Stem Cell Therapy. Blood. 2023;**142**:7129

[93] Han JP, Kim M, Choi BS, Lee JH, Lee GS, Jeong M, Lee Y, Kim EA, Oh HK, Go N, et al. In vivo delivery of CRISPR-Cas9 using lipid nanoparticles enables antithrombin gene editing for sustainable hemophilia A and B therapy. Science Advances. 2022;**8**:eabj6901

[94] Gyorgy B, Loov C, Zaborowski MP, Takeda S, Kleinstiver BP, Commins C, Kastanenka K, Mu D, Volak A,

Giedraitis V, et al. CRISPR/Cas9 Mediated Disruption of the Swedish APP Allele as a Therapeutic Approach for Early-Onset Alzheimer's Disease. Molecular Therapy - Nucleic Acids. 2018;**11**:429–440

[95] Monteys AM, Ebanks SA, Keiser MS, Davidson BL. CRISPR/Cas9 Editing of the Mutant Huntingtin Allele In Vitro and In Vivo. Molecular Therapy. 2017;**25**:12–23. DOI: 10.1016/j.ymthe.2016.11.010

[96] Levy JM, Yeh WH, Pendse N, Davis JR, Hennessey E, Butcher R, Koblan LW, Comander J, Liu Q, Liu DR. Cytosine and adenine base editing of the brain, liver, retina, heart and skeletal muscle of mice via adeno-associated viruses. Nature Biomedical Engineering. 2020;**4**:97–110. DOI: 10.1038/s41551-019-0501-5

[97] Sousa AA, Terrey M, Sakai HA, Simmons CQ, Arystarkhova E, Morsci NS, Anderson LC, Xie J, Suri-Payer F, Laux LC, et al. In vivo prime editing rescues alternating hemiplegia of childhood in mice. Cell. 2025;**188**:4275–4294.e23

[98] Maeder ML, Stefanidakis M, Wilson CJ, Baral R, Barrera LA, Bounoutas GS, Bumcrot D, Chao H, Ciulla DM, DaSilva JA, et al. Development of a gene-editing approach to restore vision loss in Leber congenital amaurosis type 10. Nature Medicine. 2019;**25**:229–233

[99] Xue Y, Tao Y, Wang X, Wang X, Shu Y, Liu Y, Kang W, Chen S, Cheng Z, Yan B, et al. RNA base editing therapy cures hearing loss induced by OTOF gene mutation. Molecular Therapy. 2023;**31**:3520–3530

[100] Li G, Jin M, Li Z, Xiao Q, Lin J, Yang D, Liu Y, Wang X, Xie L, Ying W, et al. Mini-dCas13X-mediated RNA editing restores dystrophin expression in a humanized mouse model of Duchenne muscular dystrophy. Journal of Clinical Investigation. 2023;**133**. DOI: 10.1172/JCI162809

[101] Arbab M, Matuszek Z, Kray KM, Du A, Newby GA, Blatnik AJ, Raguram A, Richter MF, Zhao KT, Levy JM, et al. Base editing rescue of spinal muscular atrophy in cells and in mice. Science. 2023;**380**:eadg6518

[102] Nishiyama T, Zhang Y, Cui M, Li H, Sanchez-Ortiz E, McAnally JR, Tan W, Kim J, Chen K, Xu L, et al. Precise genomic editing of pathogenic mutations in RBM20 rescues dilated cardiomyopathy. Science Translational Medicine. 2022;**14**:eade1633

[103] Koblan LW, Erdos MR, Wilson C, Cabral WA, Levy JM, Xiong ZM, Tavarez UL, Davison LM, Gete YG, Mao X, et al. In vivo base editing rescues Hutchinson-Gilford progeria syndrome in mice. Nature. 2021;**589**:608–614

[104] Musunuru K, Chadwick AC, Mizoguchi T, Garcia SP, DeNizio JE, Reiss CW, Wang K, Iyer S, Dutta C, Clendaniel V, et al. In vivo CRISPR base editing of PCSK9 durably lowers cholesterol in primates. Nature. 2021;**593**:429–434

[105] Gillmore JD, Gane E, Taubel J, Kao J, Fontana M, Maitland ML, Seitzer J, O'Connell D, Walsh KR, Wood K, et al. CRISPR-Cas9 In Vivo Gene Editing for Transthyretin Amyloidosis. New England Journal of Medicine. 2021;**385**:493–502

[106] Song CQ, Jiang T, Richter M, Rhym LH, Koblan LW, Zafra MP, Schatoff EM, Doman JL, Cao Y, Dow LE, et al. Adenine base editing in an adult mouse model of tyrosinaemia. Nature Biomedical Engineering. 2020;**4**:125–130

[107] Shen S, Sanchez ME, Blomenkamp K, Corcoran EM, Marco E, Yudkoff CJ, Jiang H, Teckman JH, Bumcrot D, Albright CF. Amelioration of Alpha-1 Antitrypsin Deficiency Diseases with Genome Editing in Transgenic Mice. Human Gene Therapy. 2018;**29**:861–873. DOI: 10.1089/hum.2017.227

[108] Bock D, Rothgangl T, Villiger L, Schmidheini L, Matsushita M, Mathis N, Ioannidi E, Rimann N, Grisch-Chan HM, Kreutzer S, et al. In vivo prime editing of a metabolic liver disease in mice. Science Translational Medicine. 2022;**14**:eabl9238

[109] Zabaleta N, Barberia M, Martin-Higueras C, Zapata-Linares N, Betancor I, Rodriguez S, Martinez-Turrillas R, Torella L, Vales A, Olague C, et al. CRISPR/Cas9-mediated glycolate oxidase disruption is an efficacious and safe treatment for primary hyperoxaluria type I. Nature Communications. 2018;**9**:5454

[110] Longhurst HJ, Lindsay K, Petersen RS, Fijen LM, Gurugama P, Maag D, Butler JS, Shah MY, Golden A, Xu Y, et al. CRISPR-Cas9 In Vivo Gene Editing of KLKB1 for Hereditary Angioedema. New England Journal of Medicine. 2024;**390**:432–441

[111] Greely HT. CRISPR'd babies: Human germline genome editing in the 'He Jiankui affair'. Journal of Law and the Biosciences. 2019;**6**:111–183. DOI: 10.1093/jlb/lsz010

[112] Baylis F, Darnovsky M, Hasson K, Krahn TM. Human Germ Line and Heritable Genome Editing: The Global Policy Landscape. The CRISPR Journal. 2020;**3**:365–377. DOI: 10.1089/crispr.2020.0082

Chapter 2

Gene, Cell, and RNA Therapies and Its Recent Advances: An Overview

Vasanthakumar Karumannan, Prakash Sulur Muthusamy and Akula Sindhu Bhargavi

Abstract

Gene, cell, and RNA therapies have revolutionized modern medicine, offering curative potential for genetic disorders, cancers, autoimmune diseases, and degenerative conditions. *Gene therapy* involves the delivery or modification of genetic material to correct disease-causing mutations, with recent advances in gene replacement (introducing the functional gene), gene editing (Clustered Regularly Interspaced Short Palindromic Repeats (CRISPR)/Cas9, TALENs, and Zinc Finger Nucleases (ZFNs)), gene silencing (techniques like antisense oligonucleotides (ASOs) and RNA interference (RNAi)), and viral/non-viral vector technologies. *Cell therapy* utilizes autologous or allogeneic cells to restore function, exemplified by CAR-T cell therapy, which has transformed hematologic cancer treatment, and stem-cell-based therapies are also expanding applications in regenerative medicine. *RNA-based therapies* manipulate the expression of proteins by targeting RNA molecules. They do not alter the DNA sequence but influence gene expression at the transcriptional or translational levels; for instance, *mRNA vaccines* and *small interfering RNA (siRNA)* treatments have gained prominence, particularly with the rapid development of *mRNA COVID-19 vaccines*, demonstrating their potential in infectious diseases and oncology. The approval of *patisiran*, an siRNA drug for hereditary transthyretin amyloidosis, further highlights the success of RNA-based interventions. Despite promising advances, challenges remain, including manufacturing scalability, high costs, and long-term safety concerns. Integrating *Artificial Intelligence (AI)*, *next-generation sequencing (NGS)*, and *synthetic biology* drives innovation in this field. As research progresses, gene, cell, and RNA therapies are poised to redefine therapeutic landscapes, enabling precision medicine and personalized treatments for various diseases. This chapter mainly discusses the overview of the progress so far.

Keywords: gene therapy, cell therapy, RNA therapy, CRISPR, CAR-T, mRNA vaccines, siRNA, regenerative medicine

1. Introduction

Cellular therapy is an old concept, dating back thousands of years. Paracelsus, a sixteenth-century physician, described cellular therapy as "Similus Similibus Curantur," which translates into "Likes are cured by likes" [1].

Dr. Paul Niehans, a pioneer of "Cell therapy and Regenerative Medicine" in 1931, performed an experiment to treat a critical patient, which has significant historical importance in discovering cellular therapies. When the conventional treatments failed, Dr. Niehans took a leap of faith. To save the dying patient, he injected the Ox parathyroid gland into the patient's pectoral muscles. This outcome exceeded all the established medical expectations. Not only did they recover, but they also exhibited remarkable signs of rejuvenation. This was the birth of the cellular therapies [2].

Cellular therapy (CT) is the transplantation of human cells to replace or repair damaged tissue and/or cells [3]. The terms used are cell therapy, cellular therapy, cellular immunotherapy, and regenerative medicine. They all refer to administering living cells, removing unwanted cells for therapeutic purposes, and restoring normal cellular function [1].

The currently available therapies are,

1. Gene therapy – Gene replacement, gene editing, gene silencing and vector-based therapy, and CRISPR-based therapy.

2. Cell therapy – Chimeric antigen receptor T-cell therapy (CAR-T therapy) and hematopoietic stem cell transplantation.

3. RNA-based therapies – mRNA therapy, small interfering RNA (siRNA), microRNA modulation, ribozymes, and RNA aptamers.

4. Advanced therapy medicinal products.

1.1 Gene therapy

The first successful gene therapy in humans was conducted on September 14, 1990, at the National Institutes of Health (NIH) in the United States, led by Dr. W. French Anderson, Dr. Michael Blaese, and Dr. Kenneth Culver. This groundbreaking clinical trial aimed to treat a four-year-old girl, Ashanti DeSilva, who had severe combined immunodeficiency (SCID) caused by a deficiency in the adenosine deaminase (ADA) enzyme using a retroviral vector [4].

Gene therapy uses genome-editing technology. Patients' hematopoietic (blood) stem cells are modified by gene-cutting technology. Gene therapy involves gene replacement (hemophilia), gene editing (sickle cell disease, beta thalassemia), gene silencing, and vector-based therapy.

The first FDA-approved therapy, CRISPR/Cas9, a genome-editing technology, was used to treat sickle cell disease (SCD). CRISPR/Cas9 can be directed to cut DNA in targeted areas, enabling it to accurately edit (remove, add, or replace) DNA where it was cut. The modified blood stem cells are transplanted back into the patient, where they engraft (attach and multiply) within the bone marrow and increase the production of fetal hemoglobin (HbF), which facilitates oxygen delivery [5].

Other gene therapies can treat the following conditions: Duchenne muscular dystrophy, inherited retinal disorder, cerebral adrenoleukodystrophy (CALD), and spinal muscular atrophy (SMA).

1.1.1 CRISPR-based genes

CRISPR (Clustered Regularly Interspaced Short Palindromic Repeats) gene-editing technology has revolutionized genetic medicine by enabling precise modifications in DNA sequences. CRISPR-based gene therapy utilizes the CRISPR-Cas9 system to target and edit disease-associated genes, offering a promising approach to treating genetic disorders at their root cause.

1.1.1.1 Mechanism of CRISPR gene therapy

CRISPR-Cas9 consists of two key components: the Cas9 enzyme, which acts as molecular scissors to cut DNA at a specific site, and a guide RNA (gRNA) that directs Cas9 to the target sequence. Once the DNA is cut, the cell's natural repair mechanisms—either non-homologous end joining (NHEJ) or homology-directed repair (HDR)—introduce modifications that can correct mutations, deactivate defective genes, or insert therapeutic sequences [6].

CRISPR-based therapies have shown significant promise in various fields:

- Hematologic disorders: The first FDA-approved CRISPR therapy, exagamglogene autotemcel (exa-cel), treats sickle cell disease and transfusion-dependent β-thalassemia by correcting mutations in the HBB gene [7].

- Oncology: CRISPR is being explored to engineer CAR-T cells to enhance immunotherapy for leukemia and lymphoma.

- Ophthalmic diseases: Trials using CRISPR-based therapies, such as EDIT-101, aim to treat inherited retinal disorders like Leber congenital amaurosis (LCA) [7].

- Neurodegenerative disorders: Research investigates CRISPR-based treatments for Huntington's disease and amyotrophic lateral sclerosis (ALS) by silencing pathogenic genes [7].

1.2 Cell therapy

1.2.1 Hematopoietic stem cell transplantation

Hematopoietic stem cell transplantation is also called a bone marrow transplant; this involves administering healthy stem cells into the patient's bloodstream with depleted/dysfunctional bone marrow. It mainly treats blood malignancies, leukemia, lymphoma, myelodysplastic syndrome, and myeloproliferative syndrome [8].

E. Donnall Thomas, a physician-scientist and a pioneer in bone marrow transplantation, performed the first allogeneic hematopoietic stem cell transplantation (HSCT). The physician continued his research and received a Nobel Prize for his work on stem cell transfusion in the category of Physiology and Medicine [9].

In 1957, the first transplantation was done between monozygotic twins in New York, and the first allogenic transplantation was done in 1968 to a pediatric patient with severe combined immunodeficiency in Minnesota. The field of stem cell and regenerative medicine is evolving constantly, and currently, both autologous and allogeneic HSCT are being performed worldwide [10].

HSCT can treat leukemia, lymphoma, multiple myeloma, and aplastic anemia [11]. The EBMT (The European Society for Blood and Marrow Transplantation) annual activity survey, conducted since 1990, has recorded data on nearly one million hematopoietic cell transplants (HCT). In 2023, over 47,500 transplants were performed, reflecting a recovery from the decline caused by the COVID-19 pandemic. While allogeneic HCT numbers have surpassed pre-pandemic levels, autologous HCT remains below 2019 figures. Key trends include a rise in allogeneic HCT for myeloid malignancies like AML, MDS, and MPN, and a decline in CML transplants due to effective TKI therapies. There is also a shift away from HLA-identical sibling donors toward unrelated and haploidentical donors.

Autologous HCT shows an increase for plasma cell disorders (PCD) but a decrease for lymphoma, likely due to the growing use of CAR-T cell therapy. CAR-T therapy continues to expand rapidly, particularly for non-Hodgkin lymphoma (NHL) and PCD [12].

1.2.2 CAR-T cell therapy

The first CAR-T cell therapy was successfully administered to a human in 2010 by Dr. Carl June and his team at the University of Pennsylvania. This groundbreaking therapy was used to treat a patient with relapsed chronic lymphocytic leukemia (CLL). The patient, Bill Ludwig, showed a complete remission that lasted for years, marking a significant milestone in cancer immunotherapy [13, 14].

The first pediatric patient to receive CAR-T cell therapy was Emily Whitehead, a six-year-old girl with relapsed and refractory acute lymphoblastic leukemia (ALL). In April 2012, she was treated by a team led by Dr. Carl June and Dr. Stephan Grupp at the Children's Hospital of Philadelphia (CHOP) and the University of Pennsylvania. Her case became a landmark moment in the history of CAR-T therapy and pediatric oncology [14, 15].

CAR-T cell therapy, or chimeric antigen receptor T-cell therapy, is an immunotherapy where a patient's T cells—a type of white blood cell crucial to the immune system—are engineered to better recognize and attack cancer cells. This therapy has been groundbreaking in treating certain blood cancers, such as relapsed or refractory B-cell lymphomas and leukemia [16].

CAR-T therapy currently has EMA (European Medicine Agency) approval for treating

a. B-cell acute lymphoblastic leukemia in children and young people,

b. Diffuse large B-cell lymphoma, and

c. Primary mediastinal B-cell lymphoma and multiple myeloma in adults.

CAR-T cell therapy demonstrates remarkable efficacy in treating relapsed or refractory hematologic cancers (e.g., B-cell lymphomas and acute lymphoblastic leukemia). Long-term remission rates reach 40–50% in aggressive cases, with expanding research into solid tumors showing promising but limited success due to tumor microenvironment challenges [17].

Toxicities like cytokine release syndrome (CRS) and immune effector cell-associated neurotoxicity syndrome (ICANS) remain key concerns but are increasingly manageable with improved protocols.

Cytokine Release Syndrome (CRS) results from widespread immune activation and cytokine overproduction, leading to fever, hypotension, hypoxia, and multi-organ dysfunction. It typically occurs within the first week post-infusion. ICANS is a neurotoxic complication presenting with confusion, aphasia, seizures, and, in severe cases, cerebral edema. It usually develops after CRS but can occur independently.

Infections are a major concern during CAR-T cell therapy due to profound immunosuppression caused by lymphodepleting chemotherapy and the therapy itself [18].

The management includes IL-6 inhibitors (e.g., tocilizumab) for CRS and corticosteroids for severe ICANS. However, these treatments further suppress immunity, increasing infection risks. Early identification, antimicrobial prophylaxis, and careful monitoring are essential to minimize morbidity. A multidisciplinary approach combining immunosuppressive management and infection prevention is crucial for optimizing patient outcomes during CAR-T cell therapy [19].

The currently approved CAR-T cell therapies and the approved date are mentioned below (**Table 1**).

1.3 RNA-based therapies

1.3.1 Messenger RNA (mRNA) therapy mechanism

Delivers synthetic mRNA to cells, instructing them to produce specific therapeutic proteins.

1.3.1.1 Applications

It includes the preparation of vaccines; for instance, COVID-19 vaccines use mRNA to encode the SARS-CoV-2 spike protein COVID-19 vaccine (FDA-approved, 2021) [21], cancer immunotherapy, where mRNA-based cancer vaccines stimulate the immune system to recognize and attack the tumor cells, and protein replacement therapy, where it treats enzyme deficiencies by encoding the functional proteins.

1.3.2 siRNA therapy and its applications

Patisiran is the first small interfering RNA (siRNA) drug approved by the U.S. Food and Drug Administration (FDA). It received FDA approval on August 10,

Generic name	Indications	Approval date
Tisagenlecleucel	Pediatric and young adult patients with r/r B-cell ALL; adult patients with r/r DLBCL after ≥2 lines of therapy	August 2018
Axicabtagene ciloleucel	Adult patients with r/r DLBCL and PMBCL after ≥2 lines of systemic therapy	August 2018
Brexucabtagene autoleucel	Adult patients with r/r Mantle cell lymphoma (MCL) and r/r ALL	December 2020
Lisocabtagene maraleucel	Adult patients with r/r DLBCL, high-grade B-cell lymphoma, PMBCL, and follicular lymphoma grade 3B	April 2022
Idecabtagene vicleucel	Adult patients with r/r multiple myeloma after ≥3 prior therapies	August 2021
Ciltacabtagene autoleucel	Adult patients with r/r multiple myeloma after ≥3 prior therapies	May 2022

Source: European Medicines Agency (EMA), 2025 [20]. r/r – Relapse or Refractory.

Table 1.
CAR-T therapy – Generic names, indications, and the approval date [20].

2018, for the treatment of hereditary transthyretin-mediated amyloidosis (hATTR amyloidosis) with polyneuropathy in adults. This landmark approval marked a breakthrough in RNA interference (RNAi) therapy. Patisiran uses lipid nanoparticles to deliver siRNA molecules into liver cells, where TTR is produced. The siRNA binds to and degrades the TTR mRNA, reducing both mutant and wild-type TTR protein levels in the bloodstream, thus slowing disease progression and alleviating symptoms [22].

1.3.3 miRNA: MicroRNA modulation

Modulates endogenous miRNAs to either enhance or suppress gene expression.

1.3.4 Antisense oligonucleotides (ASO)

Mechanism: Synthetic RNA strands bind to mRNA to block translation or modify splicing patterns.

1.3.4.1 Applications

It participates in SMN2 splicing modulation and skips the exon mutation to restore the dystrophin production and helps in the treatment of the following neurodegenerative conditions: Spinal Muscular Atrophy (SMA) and Duchenne Muscular Dystrophy (DMD), respectively.

The FDA-approved treatment for SMA is Nusinersen (Spinraza) (2016), and for DMD is Eteplirsen (Exondys 51) (2016) [23].

1.3.5 RNA aptamers

Mechanism: RNA molecules bind to specific targets (e.g., proteins) to inhibit their function.

1.3.5.1 Applications

RNA aptamers inhibit VEGF and prevent ARMD (Age-related Macular Degeneration). The FDA-approved treatment is Pegaptanib (Macugen), and this treats wet AMD (FDA-approved, 2004) [23]. Aptamers detect biomarkers for early cancer diagnosis as well.

1.4 Advanced therapy medicinal products

Advanced Therapy Medicinal Products (ATMPs) are biologically derived medicines that harness the power of genetic modification, cellular engineering, and tissue regeneration. These therapies are the modified products from the above-discussed therapies and have transformed modern medicine by offering curative potential in inherited genetic disorders, malignancies, and degenerative diseases. The European Medicines Agency (EMA) and the U.S. Food and Drug Administration (FDA) regulate ATMPs under stringent guidelines to ensure safety and efficacy.

1.4.1 Definition and classification

According to the EMA, ATMPs are categorized into four major types:

1. Gene Therapy Medicinal Products (GTMPs) – Involve the insertion, deletion, or modification of genetic material to treat diseases at a molecular level. For instance, Luxturna (voretigene neparvovec) is an FDA-approved gene therapy for inherited retinal dystrophy caused by RPE65 mutations.

2. Somatic-Cell Therapy Medicinal Products (sCTMPs) – Utilize manipulated cells to repair or replace damaged tissue. Somatic-cell therapy involves the use of autologous or allogeneic cells for therapeutic benefit. CAR-T cell therapy (Chimeric Antigen Receptor T-cell therapy) has been a breakthrough in oncology, with drugs such as Kymriah (tisagenlecleucel) and Yescarta (axicabtagene ciloleucel) demonstrating high efficacy in hematological malignancies [24].

3. Tissue-Engineered Products (TEPs) – Comprise cells or scaffolds engineered to restore, maintain, or improve tissue function. Tissue-engineered products restore damaged tissues using scaffolds and autologous cells. Holoclar, a stem-cell-based treatment, restores corneal epithelium in patients with limbal stem cell deficiency (LSCD) [24].

4. Combined ATMPs – Contain ATMP components along with medical treatments.

The currently approved therapies are available on the FDA website as of 2024 (**Table 2**) [25]:

Drug name	Indication (Disease treated)
Idecabtagene vicleucel	Relapsed/refractory multiple myeloma
Lifileucel	Unresectable/metastatic melanoma
Nadofaragene firadenovec-vncg	High-risk BCG-unresponsive non-muscle invasive bladder cancer
HPC, Cord blood	Hematopoietic progenitor cell transplantation
Obecabtagene autoleucel	Relapsed/refractory B-cell precursor acute lymphoblastic leukemia (ALL)
Fidanacogene elaparvovec-dzkt	Moderate to severe hemophilia B
Lisocabtagene maraleucel	Large B-cell lymphoma, chronic lymphocytic leukemia (CLL), small lymphocytic lymphoma (SLL), follicular lymphoma, mantle cell lymphoma
Ciltacabtagene autoleucel	Relapsed/refractory multiple myeloma
Exagamglogene autotemcel	Sickle cell disease (SCD), transfusion-dependent ß-thalassemia (TDT)
Delandistrogene moxeparvovec-rokl	Duchenne muscular dystrophy (DMD)
Allogeneic cultured keratinocytes and fibroblasts	Mucogingival conditions
Etranacogene dezaparvovec-drlb	Hemophilia B

Drug name	Indication (Disease treated)
Talimogene laherparepvec	Unresectable recurrent melanoma
Eladocagene exuparvovec-tneq	Aromatic L-amino acid decarboxylase (AADC) deficiency
Tisagenlecleucel	B-cell precursor ALL, large B-cell lymphoma, follicular lymphoma
Donislecel-Jujn	Type 1 diabetes with severe hypoglycemia
Azficel-T	Nasolabial fold wrinkles
Atidarsagene autotemcel	Metachromatic leukodystrophy (MLD)
Voretigene neparvovec-rzyl	RPE65 mutation-associated retinal dystrophy
Lovotibeglogene autotemcel	Sickle cell disease (SCD)
Autologous cultured chondrocytes on porcine collagen membrane	Knee cartilage defects
Omidubicel-only	Hematologic malignancies (for umbilical cord blood transplantation)
Sipuleucel-T	Metastatic castrate-resistant prostate cancer
Allogeneic processed thymus tissue–agdc	Immune reconstitution in congenital athymia
Valoctocogene roxaparvovec-rvox	Severe hemophilia A
Remestemcel-L-rknd	Steroid-refractory acute graft-versus-host disease (SR-aGvHD) in pediatric patients
Acellular tissue engineered vessel-tyod	Extremity arterial injury
Elivaldogene autotemcel	Cerebral adrenoleukodystrophy (CALD)
Allogeneic cultured keratinocytes and dermal fibroblasts in murine collagen-dsat	Debrided thermal burns
Brexucabtagene autoleucel	Mantle cell lymphoma (MCL), B-cell precursor ALL
Afamitresgene autoleucel	Unresectable/metastatic synovial sarcoma
Beremagene geperpavec	Dystrophic epidermolysis bullosa
Axicabtagene ciloleucel	Large B-cell lymphoma
Betibeglogene autotemcel	ß-thalassemia requiring regular RBC transfusions
Onasemnogene abeparvovec-xioi	Spinal muscular atrophy (SMA)

Table 2.
FDA-approved advanced therapy medicinal products (as of December 20, 2024) [25].

2. Conclusion

Gene, cell, and RNA therapies have ushered in a transformative medical era, offering groundbreaking treatment modalities for genetic disorders, malignancies, autoimmune conditions, and degenerative diseases. These advanced therapies rapidly redefine the therapeutic landscape by addressing disease pathophysiology at a molecular and cellular level, moving beyond symptomatic relief to potential cures.

With the advent of CRISPR-Cas9 and other genome-editing tools, gene therapy has demonstrated remarkable precision in correcting genetic abnormalities. Approving

CRISPR-based treatments, such as exagamglogene autotemcel (exa-cel) for sickle cell disease, marks a significant milestone in personalized medicine. Similarly, cell-based therapies, including hematopoietic stem cell transplantation (HSCT) and chimeric antigen receptor T-cell (CAR-T) therapy, have revolutionized the management of hematologic malignancies, improving survival rates for patients with relapsed or refractory cancers. Additionally, regenerative medicine approaches utilizing stem cells hold immense promise for tissue repair and organ regeneration [6].

RNA-based therapies, particularly mRNA vaccines, have played a pivotal role in controlling the COVID-19 pandemic, showcasing their potential for rapid therapeutic development. Beyond infectious diseases, RNA therapeutics, such as small interfering RNA (siRNA) and antisense oligonucleotides, are advancing treatments for rare genetic disorders, neurodegenerative conditions, and oncology. The success of patisiran, the first FDA-approved siRNA drug for hereditary transthyretin amyloidosis, underscores the clinical potential of RNA technologies [25].

Despite these advances, several challenges persist. Manufacturing scalability, high production costs, and the complexities of regulatory approval limit the widespread adoption of these therapies. Concerns regarding long-term safety, immune responses, and off-target effects necessitate continuous research and long-term follow-up studies. Ethical considerations, particularly in gene editing and germline modifications, require stringent oversight to balance innovation with responsible medical practice.

The future of gene, cell, and RNA therapies lies in their integration with emerging technologies such as Artificial Intelligence (AI), next-generation sequencing (NGS), and synthetic biology. These innovations refine therapeutic targeting, improve efficiency, and enhance safety profiles. Developing off-the-shelf cell therapies, improved gene delivery vectors, and personalized RNA-based interventions will further expand treatment options. As research advances, regulatory bodies and healthcare systems must collaborate to ensure equitable access to these life-changing therapies [24].

In conclusion, gene, cell, and RNA therapies represent a paradigm shift in modern medicine, offering personalized and curative solutions for complex diseases. Continued innovation, ethical oversight, and global collaboration will be essential in harnessing the full potential of these groundbreaking therapeutic strategies, ultimately shaping the future of precision medicine.

Acknowledgements

I am grateful to *Dr. Iain Wilkinson*, *Dr. Ben Mearns, and the Surrey ward* for their unwavering support in my clinical role. Their encouragement has been instrumental in helping me strike a balance between my clinical and academic responsibilities. I am profoundly grateful to *East Surrey Hospital* for shaping my early career and providing invaluable opportunities that continue to pave the way for my professional growth.

I sincerely thank *Professor Simon Maxwell* for inspiring me to pursue academic pursuits at the *University of Edinburgh*. Professor, your mentorship has been pivotal in broadening my academic horizons, and I am truly honored to be one of your students.

I also sincerely appreciate *Dr. Lakhvinder Singh's* steadfast support and invaluable guidance throughout my academic journey at *PGIMER*. Your wisdom, mentorship, and principles have been a constant source of motivation, and I am sincerely grateful for your belief in my potential.

Thank you to all those who have supported and encouraged me along this journey. Your influence has played a significant role in shaping my aspirations and achievements.

Conflict of interest

The authors declare no conflict of interest.

Other declarations

All the references have been quoted appropriately.

Considering the difficulty of the topic, artificial intelligence is used minimally to simplify the complexity of the subject. However, the originality of the topic is preserved. The objective of sharing knowledge about the overview of therapies is achieved.

Author details

Vasanthakumar Karumannan[1*], Prakash Sulur Muthusamy[2] and Akula Sindhu Bhargavi[3]

1 East Surrey Hospital, Surrey and Sussex NHS Healthcare Trust, Surrey, United Kingdom

2 University Hospitals of Coventry and Warwickshire NHS Trust, Coventry, United Kingdom

3 Department of Transfusion Medicine, All India Institute of Medical Sciences (AIIMS), Mangalagiri, Andra Pradesh, India

*Address all correspondence to: v.karumannan@nhs.net

References

[1] Atilla E, Kilic P, Gurman G. Cellular therapies: Day by day, all the way. Transfusion and Apheresis Science. 2018;**57**(2):187-196

[2] Abgrall JM. Healing or Stealing?: Medical Charlatans in the New Age. New York: Algora Pub; 2000

[3] O'Connor DM. Introduction to Gene and Stem-Cell Therapy. London: Academic Press is an Imprint of Elsevier; 2017. pp. 141-165

[4] Blaese RM, Culver KW, Miller AD, Carter CS, Fleisher T, Clerici M, et al. T lymphocyte-directed gene therapy for ADA – SCID: Initial trial results after 4 years. Science. 1995;**270**(5235):475-480

[5] Cappelli B, Gluckman E, Corbacioglu S, de la Fuente J, Abboud MR. Hemoglobinopathies (sickle cell disease and thalassemia). In: Sureda A, Corbacioglu S, Greco R, Kröger N, Carreras E, editors. The EBMT Handbook: Hematopoietic Cell Transplantation and Cellular Therapies. Cham (CH): Springer, Copyright 2024, The Author(s); 2024. pp. 725-739

[6] Lee H, Rho W-Y, Kim Y-H, Chang H, Jun B-H. CRISPR-Cas9 gene therapy: Non-viral delivery and stimuli-responsive Nanoformulations. Molecules. 2025;**30**(3):542

[7] Uddin F, Rudin CM, Sen T. CRISPR gene therapy: Applications, limitations, and implications for the future. Frontiers in Oncology. 2020;**10**:2-12

[8] Khaddour K, Hana CK, Mewawalla P. Hematopoietic Stem Cell Transplantation. StatPearls. Treasure Island (FL): StatPearls Publishing Copyright © 2024: StatPearls Publishing LLC; 2024

[9] Khaddour K, Hana CK, Mewawalla P. Hematopoietic Stem Cell Transplantation. StatPearls. Treasure Island (FL): StatPearls Publishing, Copyright © 2025; 2025

[10] Blau CA. E. Donnall Thomas, M.D. (1920-2012). Stem Cells Translational Medicine. 2013;**2**(2):81-82

[11] Passweg JR, Baldomero H, Ciceri F, Corbacioglu S, de la Cámara R, Dolstra H, et al. Hematopoietic cell transplantation and cellular therapies in Europe 2021. The second year of the SARS-CoV-2 pandemic. A report from the EBMT activity survey. Bone Marrow Transplantation. 2023;**58**(6):647-658

[12] Passweg JR, Baldomero H, Atlija M, Kleovoulou I, Witaszek A, Alexander T, et al. The 2023 EBMT report on hematopoietic cell transplantation and cellular therapies. Increased use of allogeneic HCT for myeloid malignancies and of CAR-T at the expense of autologous HCT. Bone Marrow Transplantation. 2025:02-09

[13] Khan I, Khan N, Wolfson N, Djebabria K, Rehman MEU, Anwer F. Safety of CAR-T cell therapy in patients with renal failure/acute kidney injury: Focused review. Clinical Hematology International. 2023;**5**(2-3):122-129

[14] Mitra A, Barua A, Huang L, Ganguly S, Feng Q, He B. From bench t bedside: The history and progress of CAR T cell therapy. Frontiers in Immunology. 2023;**14**:1-14

[15] Grupp SA, Kalos M, Barrett D, Aplenc R, Porter DL, Rheingold SR, et al. Chimeric antigen receptor–modified T cells for acute lymphoid Leukemia. New

England Journal of Medicine. 2013;**368**(16):1509-1518

[16] Board PDQPTE. Pediatric Chimeric Antigen Receptor (CAR) T-Cell Therapy (PDQ®): Health Professional Version. PDQ Cancer Information Summaries. Bethesda (MD): National Cancer Institute (US); 2002

[17] Neelapu SS, Locke FL, Bartlett NL, Lekakis LJ, Miklos DB, Jacobson CA, et al. Axicabtagene Ciloleucel CAR T-cell therapy in refractory large B-cell lymphoma. New England Journal of Medicine. 2017;**377**(26):2531-2544

[18] Santomasso BD, Park JH, Salloum D, Riviere I, Flynn J, Mead E, et al. Clinical and biological correlates of neurotoxicity associated with CAR T-cell therapy in patients with B-cell acute lymphoblastic Leukemia. Cancer Discovery. 2018;**8**(8):958-971

[19] Sterner RC, Sterner RM. CAR-T cell therapy: Current limitations and potential strategies. Blood Cancer Journal. 2021;**11**(4):2-11

[20] Agency EM. CAR T-Cell Therapies: Approved Indications and Product Information Amsterdam 2025. Available from: https://www.ema.europa.eu [Accessed: March 13, 2025]

[21] Qin S, Tang X, Chen Y, Chen K, Fan N, Xiao W, et al. mRNA-based therapeutics: Powerful and versatile tools to combat diseases. Signal Transduction and Targeted Therapy. 2022;**7**(1):2-8

[22] Adams D, Gonzalez-Duarte A, O'Riordan WD, Yang C-C, Ueda M, Kristen AV, et al. Patisiran, an RNAi therapeutic, for hereditary transthyretin amyloidosis. New England Journal of Medicine. 2018;**379**(1):11-21

[23] Zhang C, Zhang B. RNA therapeutics: Updates and future potential. Science China Life Sciences. 2023;**66**(1):12-30

[24] Hanna E, Rémuzat C, Auquier P, Toumi M. Advanced therapy medicinal products: Current and future perspectives. Journal of Market Access & Health Policy. 2016;**4**:1-8

[25] Administration USFaD. Approved Cellular and Gene Therapy Products [Internet]: FDA; 2024. Available from: https://www.fda.gov/ [Accessed: February 23, 2025]

Section 2

DNA Integrity, Repair and Aging

Chapter 3

DNA Damage, DNA Repair, and Epigenetic Alterations are Central to Aging

Carol Bernstein and Harris Bernstein

Abstract

The concept that unrepaired DNA damages are the underlying cause of aging has been supported by substantial evidence since it was first proposed in 1967. Aging is a broadly general characteristic of eukaryotes, including single-celled organisms such as paramecia. However, aging has mainly been studied in mammals. The two major causes of aging appear to be unrepaired DNA damage and epigenetic methylations or demethylations of cytosines that remain after the repair of DNA damage. Numerous studies have described increases in DNA damage with age in a variety of mammalian cell types (e.g., brain, muscle, liver, kidney, oocytes, and stem cells). In humans, several inherited genetic disorders caused by deficient repair of DNA damages are associated with premature aging. Expression of the approximately 20,000 genes in the human genome is controlled by genomic sequences referred to as enhancers and promoters. Enhancers and promoters are subject to epigenetic modifications. In promoter DNA, epigenetic alterations include methylations or demethylations of cytosines within CpG islands. With respect to enhancer DNA, epigenetic methylations or demethylations of transcription factor binding sites can alter enhancer activity. DNA repair activity often leaves behind epigenetic alterations that repress or amplify the activities of enhancer and promoter sequences. Epigenetic modifications due to DNA repair accumulate with age, thus contributing to the deficits seen as aging. Thus, aging at the cellular level is apparently due to both the accumulation of unrepaired DNA damages and the processes that repair DNA but result in lasting detrimental epigenetic alterations of gene expression.

Keywords: DNA damage, DNA repair, epigenetics, aging, genome, gene expression, cytosine methylation, enhancer, promoter

1. Introduction

Although research on aging has mainly focused on humans and other mammalian species, aging has also been demonstrated and characterized in single-celled organisms such as *Paramecium tetraurelia* [1], yeast [2], and even the bacterium *Escherichia coli* [3]. On the basis of our current state of knowledge, we propose that aging is a broadly general process in cellular organisms and that there are two main causes of aging at the molecular level. These are (1) unrepaired damage in DNA that accumulates over time and (2) changes in how genes are expressed subsequent to DNA repair, due to modifications of DNA that alter gene expression (epigenetic changes).

Organisms with a DNA genome are susceptible to DNA damage [4]. Ordinarily, damage to DNA is accurately repaired shortly after its occurrence, resulting in restored chromosome structure and genome sequence. However, a small fraction of the damage that occurs remains unrepaired or is repaired inaccurately, resulting in the decay of appropriate gene expression over time. We consider that this process of decay is the fundamental basis of aging.

Aging has been well studied in humans and in several different mammalian species. The idea that aging is due to DNA damage (as distinct from mutation) was first proposed in 1967 [5] and more fully developed with supporting evidence in 1985 [6]. Further evidence supporting the DNA damage theory of aging was presented in the 1990s [7–11]. In more recent years, further evidence has emerged that decreased repair of DNA damage has a causal role in aging [12, 13].

Most recently, evidence has been emerging that, although the repair of DNA may accurately regenerate the original nucleotide sequence, other aspects of the damaged chromosome region retain scars of the original damage. These scars, referred to as epigenetic alterations, also contribute to aging. Evidence that altered or lost epigenetic information is a primary cause of aging has been reviewed in several recent publications [14–16]. Epigenetic changes often occur in promoter and enhancer DNA sequences, and their accumulation over time appears to contribute to cellular aging. The expression of the approximately 20,000 genes in the human genome is controlled by enhancers and promoters. As discussed below, epigenetic alterations include methylations or demethylations of cytosine in CpG islands in promoter DNA sequences. Additionally, methylations or demethylations of transcription factor binding sequences within enhancers can alter their activity.

2. DNA damage: A primary cause of aging

Considerable evidence indicates that DNA damage can affect almost all aspects of the aging process, so that it is potentially a unifying cause of aging [17]. DNA damages occur frequently in somatic cells, amounting to about 70,000 DNA damages per day in humans [18]. Most damages are repaired, but those damages that are not repaired tend to interfere with gene transcription and DNA replication. Accumulation of such interference can cause progressive impairments of cellular function and may lead to cell death. Mammalian cells that are long-lived, such as neurons, differentiated muscle cells, and other types of differentiated cells, do not divide or only slowly divide and tend to accumulate DNA damage with age. The decline in function of such cell types appears to be a major factor in mammalian aging.

The association of DNA damage with aging has been widely studied, and the aging effects of DNA damage have been reported to occur in numerous organs of mammals. We next briefly review evidence of DNA damage accumulation in aging cells of the mammalian brain, muscle, oocytes, hematopoietic stem cells, liver, and kidney.

2.1 Increase in DNA damage with age in the brain

At least 29 separate reports have shown an increase in DNA damage in the mammalian brain with aging [19]. The mammalian species in which DNA damage was found to increase in the brain with age include mouse, rat, gerbil, rabbit, dog, and humans; and the types of damage that occur include single- and double-strand breaks, oxidative modification of the guanine base in DNA (8-OHdG), and other types of damage [19]. In these studies, a wide variety of different techniques were used to measure DNA damage. As an example of these studies, one study used the comet assay to measure DNA damage in neurons and astrocytes from the cortex of young (7 days old), adult (6 months old), and old (2 years old) rats [20]. A considerable increase was observed in damaged DNA (single- and double-strand breaks) in both neurons and astrocytes by 6 months of age, and there was a further increase in damage by 2 years of age [20]. Recent innovative approaches over the past decade indicate that damage to DNA is a major risk factor for the decrease in neuronal functions with age, and these approaches have begun to reveal the characteristics of neuronal DNA damage hot spots and the causes of DNA damage [21].

2.2 Increase in DNA damage with age in muscle

Increases in DNA damage in mammalian muscle with age (particularly the heart and skeletal muscle) were shown in at least 18 studies [19]. The mammalian species in which DNA damage was found to increase in muscle with age included mouse, rat, gerbil, and humans; and the types of damage that occurred included single-strand breaks, oxidative modification of the guanine base in DNA (8-OHdG), and DNA–protein crosslinks [19]. In one study, as an example, an age-dependent exponential increase in 8-OHdG in heart DNA was observed in the rat [22]. Also, in this study, there was an age-dependent increase in DNA–protein crosslinks and DNA double-strand and single-strand breaks [22].

2.3 Accumulation of DNA damage with age in oocytes

In humans, oocyte loss accelerates as women approach their mid-30s. Oocytes accumulate DNA damage as maternal age increases, and this appears to be a primary cause of oocyte loss and loss of reproductive potential [23]. The accumulation of DNA damage in oocytes with age reflects a reduced capability to accurately repair DNA, as well as a shift to error-prone DNA repair pathways [23].

2.4 Deficient repair of DNA damage with age in stem cells

Multipotent stem cells that are tissue-specific produce differentiated effector cells through a sequence of increasingly committed progenitor intermediates. This differentiation process is most well understood in the hematopoietic system. Hematopoiesis begins with long-term hematopoietic stem cells that can self-renew and also produce progenitor cells that, after further replication, proceed through a series of stages leading to differentiated cells lacking the capacity to self-renew. In mice, deficiencies in the ability to repair DNA limit the capacity of hematopoietic stem cells to function and to undergo self-renewal with age [24]. Hematopoietic stem cells appear to undergo intrinsic aging, and this may be caused, at least in part, by DNA damage [25].

2.5 DNA damage increases with age in the liver

Hepatocytes ordinarily do not divide. However, they retain the capacity to divide and form new cells when the liver suffers injury. Many studies have indicated that DNA damage in rodent liver increases with age [19]. The steady-state level of the oxidatively damaged DNA base guanine (8-oxo-deoxyguanosine and 8-oxo-guanine) occurred in young rat liver at about 24,000 per cell, while in old rat liver, it was about 66,000 per cell [26].

2.6 DNA damage increases with age in the kidney

Aging in the kidney is shown by reduced renal blood flow and glomerular filtration rate, as well as by a lowered ability to concentrate urine and conserve sodium and water. There have been a number of studies showing an increase in DNA damage in the kidney with age, particularly oxidative DNA damage. For example, when the level of the oxidative DNA damage 8-OHdG was measured in the kidneys of young rats (9-week-old) and aged rats (20-month-old), this damage was found to accumulate with age [27].

2.7 Genetic syndromes with defective DNA repair and premature aging

In humans, several rare inherited disorders with DNA repair defects show multiple striking features of premature aging, consistent with the concept that DNA damage is the underlying cause of aging. These syndromes include Werner syndrome (mean lifespan of 47 years), Hutchinson–Gilford progeria syndrome (HGPS) (mean lifespan of 13 years), and Cockayne syndrome (mean lifespan of 13 years).

Werner syndrome is associated with multiple features of premature aging and is caused by a genetic defect in a particular enzyme (with both DNA helicase and exonuclease activity) that functions in the DNA base excision repair pathway [28].

Hutchinson–Gilford progeria syndrome is a premature aging condition in which individuals die in their teens from myocardial infarction or stroke. HGPS is caused by a defective lamin A protein that is employed in the repair of DNA double-strand breaks [29].

Cockayne syndrome is characterized by aging traits such as growth failure, neurological degeneration, and cataracts [30]. The inability to repair oxidative base damage (such as 5-hydroxyguanine) in transcriptionally active genes through the process of transcription-coupled nucleotide excision repair appears to be the underlying cause of Cockayne syndrome [30, 31].

The studies of these three inherited conditions, as well as studies of several additional rare inherited premature aging conditions in humans, have tended to link limitations in DNA repair with accelerated aging.

3. Mutation: An unlikely primary cause of aging in mammals

Mutations, as distinct from DNA damages, are alterations in the sequence of base pairs of DNA, but the bases themselves are structurally unaltered. DNA damages, on the other hand, include chemical alterations of the normal DNA bases, as well as DNA strand breaks. DNA damages, in contrast to mutations, cannot be replicated and, therefore, cannot be inherited. In addition, DNA damages, unlike mutations, can be repaired. Accumulation of mutations, as distinct from DNA damages, has also been proposed as a cause of aging [32]. However, in diploid cells, with two copies of each gene, most gene mutations are recessive to the corresponding wild-type allele of the gene and thus do not ordinarily express.

The gene *Pms2* encodes an enzyme that corrects base mispairs in DNA [33]. Mice with a genetic mutation in the *Pms2* gene exhibit roughly a 100-fold increase in mutation frequency in all tissues examined, but apparently do not experience accelerated aging [33]. This finding, along with similar related findings [34–36], suggests that mutations are not a primary cause of aging in mice, a diploid organism. Nevertheless, mutations (in addition to DNA damage and epigenetic alterations) may contribute to aging in some species, particularly in those that have an extended haploid phase.

Although general skepticism has been expressed that mutations in the genome of the cell nucleus are the basis of aging in mouse models, a model in which aging is due specifically to mutations in mitochondrial DNA has been considered a reasonable alternative possibility. It has been reported that mutations build up in mitochondrial DNA with age in infrequently replicating cells [37]. Mitochondrial DNA is replicated by DNA polymerase gamma. A mutant mouse defective in DNA polymerase gamma can only replicate its mitochondrial DNA inaccurately, resulting in a 500-fold higher level of mitochondrial DNA mutation compared to the level of mitochondrial mutation in normal mice. These mutant mice, however, do not display rapidly accelerated aging [37]. In general, research findings do not support the idea that mutations are the primary cause of aging in mammals.

In other experiments, it was found that $Xpd^{m/m}$ mice, defective in a particular pathway for repairing DNA damage, clearly show premature aging but do not have an elevated rate of mutation [38]. This is consistent with the idea that DNA damage (as distinct from mutation) does cause aging.

4. Aging in single-celled organisms

Aging has been demonstrated in several different single-celled organisms. In these organisms, clonal reproduction is accompanied by a gradual loss of vitality over time or through successive mitotic cell divisions.

4.1 Paramecium tetraurelia

One well-studied example of clonal aging is *Paramecium tetraurelia*. During asexual reproduction, the paramecia divide by mitosis but gradually lose vitality (aging) and expire after about 200 cell divisions unless the paramecia undergo a sexual process that involves meiosis (in this case, either autogamy or conjugation). It was demonstrated experimentally that the basis for clonal aging in these paramecia resides in the macronucleus rather than the cytoplasm of the protozoa [39]. Research from several laboratories [1, 40, 41] further demonstrated that DNA damage increases dramatically during clonal aging and is the likely cause of senescence. Thus, in the case of paramecia, the accumulation of DNA damage in the macronucleus of clonally dividing cells over successive asexual generations appears to be basic to aging. This form of aging appears to be similar, at the cellular level, to aging in multicellular organisms in the sense that aging is due to the accumulation of DNA damage over time.

4.2 Saccharomyces cerevisiae

Nonreplicating cells of the yeast *Saccharomyces cerevisiae*, when starved, accumulate DNA damages such as apurinic/apyrimidinic sites and double-strand breaks [42]. The ability to repair such double-strand breaks declines during chronological aging [2]. These findings are consistent with the idea that aging is due to a reduced ability to cope with DNA damage over time.

4.3 *Escherichia coli*

When the bacterium *Escherichia coli* reproduces by binary fission, the two cells formed by cell division are functionally asymmetric, since it can be demonstrated that one new cell (the old pole cell) acts as an aging parent that repeatedly forms rejuvenated offspring [3]. On the basis of this finding, it was suggested that no life strategy is immune to the effects of aging [3]. When *E. coli* are subjected to an increased level of stress, the accumulation of damage in the old *E. coli* lineage may be greater than its immortality threshold, so that cell division is arrested and the cells become mortal [43]. Cellular aging thus appears to be a general process that affects both prokaryotes and eukaryotes [43].

5. Aging in plants

5.1 Seed senescence

A major determinant of crop yield is how well seeds germinate. The quality of plant seeds is reduced with age, and this reduced quality is associated with the accumulation of DNA damage [44]. The seeds of the rye plant undergo aging that is accompanied by the accumulation of DNA damage and loss of viability in embryos [45]. Dry seeds of the broad bean plant *Vicia faba* accumulate DNA damage over time when stored, but these DNA damages can be repaired when they germinate [46]. In the flowering plant *Arabidopsis,* during seed germination, a DNA ligase carries out the repair of DNA single- and double-strand breaks, and this ligase activity is a key determinant of seed longevity [47]. The ATM protein kinase, an enzyme involved in the cellular response to DNA damage, appears to play a major role in controlling the germination of aged seeds by integrating the DNA repair response to DNA damages accumulated during the dry state with progression through germination [48].

5.2 Tree longevity

The enzyme poly(ADP-ribose) polymerase (PARP) plays a key role in the repair of DNA damage. In trees, genes that encode PARP show significant expansion in copy number compared to annual and perennial herbs [49]. This finding suggests that the slower aging and relatively long lifespan of trees are related to the repair of DNA damage [49].

6. Alteration of epigenetic information as a primary cause of aging

6.1 Epigenetic alterations

Epigenetic alterations can cause upregulation or repression of gene expression without changing the sequence of DNA. Three main epigenetic alterations have been identified: (1) DNA methylation or demethylation of DNA, (2) alteration of histones, and (3) altered expression of noncoding RNAs, such as microRNAs [50]. In terms of aging, the most well-studied alteration has been methylation or demethylation of DNA (see **Figure 1** for methylation of the base cytosine). The age of an individual can be predicted by an epigenetic clock based on DNA methylation [51]. As detailed below, permanently altered methylations of DNA can occur during the repair of oxidative damage to DNA or during the repair of double-strand breaks in DNA.

6.2 Promoters and enhancers have specific transcription factor binding motifs that change with age

In different tissues, tens to thousands of CpG dinucleotides in promoters and enhancers have altered cytosine methylation with age [52]. Promoters and enhancers, with their enhancer-bound specific transcription factors, determine when and to what extent genes are expressed [53] (**Figure 2**). The cytosine methylation alterations to enhancers occur at the DNA sequence motif binding sites of transcription factors. The transcription factors bound to the enhancer transmit their control to the promoter through interactions with mediator proteins [54]. The core promoter is usually defined as a region of about 50 base pairs upstream and 50 base pairs downstream of the transcription start site of a gene. The core promoter, with its bound general transcription factors, is also a binding site for RNA polymerase II (RNAPII). A promoter, with RNAPII at its site, can initiate the transcription of an associated protein-coding gene, but such initiations by themselves mostly have low basal activity. However, promoters bound to general transcription factors and also associated with active enhancers bound to multiple specific transcription factors, as illustrated in **Figure 2**, can initiate higher levels of expression of their associated genes [55].

6.3 Cytosine methylation or demethylation

In humans, cytosine methylation (see **Figure 1**) tends to increase with age in the promoters of many genes, whereas cytosine methylation tends to decrease with age in many enhancers [52] (**Figure 2**). Cytosine methylation or demethylation in the

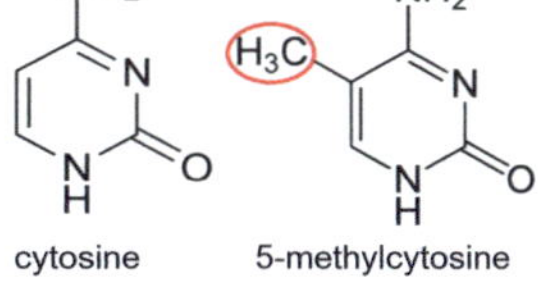

Figure 1.
Cytosine and 5-methylcytosine.

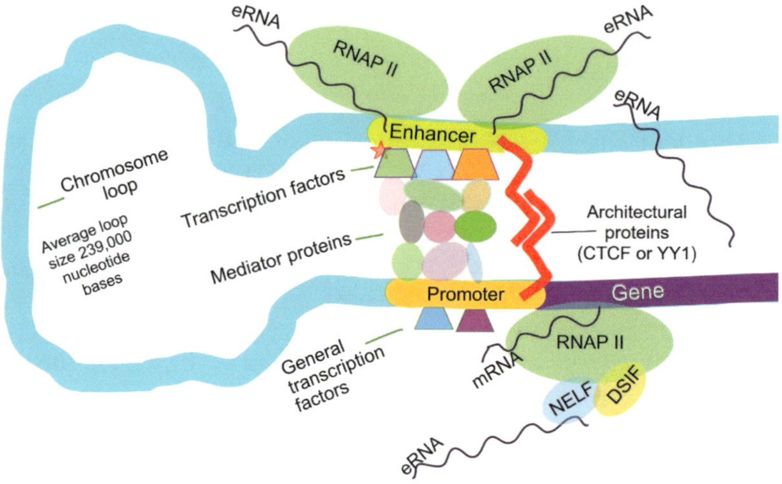

Figure 2.
Enhancers in regulation of transcription in mammals.An active enhancer regulatory sequence of DNA is enabled to interact with the promoter DNA regulatory sequence of its target gene by formation of a chromosome loop. This can initiate messenger RNA (mRNA) synthesis by RNA polymerase II (RNAP II). The loop is stabilized by one architectural protein anchored to the enhancer and one anchored to the promoter and these proteins are joined to form a dimer (red zigzags). Specific regulatory transcription factors bind to DNA sequence motifs on the enhancer. General transcription factors bind to DNA sequence motifs on the Promoter. When a transcription factors is activated by a signal (here indicated as phosphorylation shown by a small red star on a transcription factor on the enhancer) the enhancer is activated and can now activate its target promoter. The active enhancer is transcribed on each strand of DNA in opposite directions by bound RNAP IIs. Mediator (a complex consisting of about 26 proteins in an interacting structure) communicates regulatory signals from the enhancer DNA bound transcription factors to the promoter.

DNA of differentiated cells occurs when cytosine is adjacent to guanine in the sequence 5'-cytosine-phosphate-3'-guanine (called the CpG sequence). In humans, 70–80% of cytosines are methylated at CpG sites [56]. The methylation or demethylation of cytosine at CpG dinucleotides is important in the regulation of human genome architecture and activity [57]. Jain et al. [52], examining nine types of tissues (breast, lung, colon, ovary, prostate, skeletal muscle, testis, whole blood, and kidney), found 162,002 unique hypermethylated and 90,626 hypomethylated CpG sites in humans that differed with age in different tissues. Skeletal muscle had just tens of sites of altered cytosine methylation with age, while other tissues had thousands of altered methylations with age. They also found that hypermethylated CpG sites tended to occur in CpG islands within the promoters of genes, while hypomethylated CpG sites tended to occur in enhancers.

In a Chinese study of whole blood cells with age in humans, Zheng et al. [58] identified 331 positions of the human genome where methylation of cytosine increased with age and 1,330 positions where cytosine methylation declined with age. As in the studies of Jain et al. [52], cited above, Zheng et al. [58] found that the occurrences of upregulated cytosine methylation with age were in CpG islands within the promoters of genes, while occurrences of downregulated methylated cytosines with age were in the "open sea" areas (5,000 to 500,000 nucleotides away from a gene being regulated, where enhancers occur).

DOI: http://dx.doi.org/10.5772/intechopen.1012704

6.4 Timing of methylation of cytosines during development and alteration of cytosine methylation after the repair of oxidative DNA damage

In mouse reproduction (similar to human reproduction), the newly fertilized egg (the zygote) actively removes almost all methyl groups from cytosines of male origin within 6 hours [59], and during early replications, methyl groups are not added to the cytosines of female origin [59]. So, almost no methylation of cytosines is present in the embryonic morula (16-cell stage) before cells start to undergo differentiation (see **Figure 3**). As cells differentiate into tissues, methyl groups are added to cytosines as epigenetic control elements (**Figure 3**). The added cytosines determine which genes are expressed in the differentiated cells. The pattern of cytosine methylation in differentiated cells determines, to a large extent, their cellular function. Differentiation-programmed methylation of cytosines in differentiated somatic cells of the human genome occurs 99.98% of the time at CpG dinucleotides [60].

CpG dinucleotide regions of the genome tend to have their guanines preferentially targeted by oxidants due to CpG regions having low oxidation-reduction potential [61]. When the cytosine in the CpG dinucleotide is methylated (MeCpG), the guanines have an even lower oxidation-reduction potential, making these guanines even more likely to be oxidized [61]. Oxidation of the guanine in a MeCpG dinucleotide to create an 8-oxo-2'-deoxyguanosine (8-oxo-dG or 8-OHdG) can promote demethylation of the methyl group from the associated cytosine [62]. As discussed in more detail below, MeCpG dinucleotides, with the methyl group either added or removed within promoters and enhancers, can repress or activate gene transcription. Repressing or activating transcription of genes changes the programming required for cell-type function and tends to disrupt the function of cells. Loss of cellular function is a central feature of aging [63]. There are approximately 19,500 protein-coding genes in the human genome [64]. Even small changes in the expression of several hundred or a few thousand genes per cell with age can cause many cells to have disrupted cell function, a characteristic of aging.

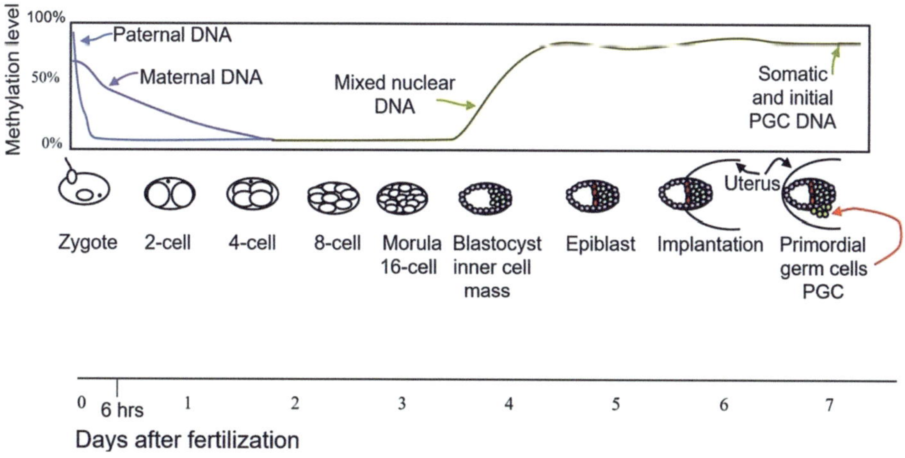

Figure 3.
methylation levels during mouse embryonic development Creative Commons Attribution License.(http://creativecommons.org/licenses/by/3.o) Carol Bernstein

6.5 Transcription factor binding motifs

In the DNA methylation aging clock of Zheng et al. [58], demethylation of transcription factor binding site motifs was found to be associated with increased expression of the associated genes. Transcription factors are proteins with short, specific amino acid sequences within their structure that allow them to bind to particular DNA sequences. The DNA sequences, known as transcription factor binding motifs, are usually 6–20 base pairs long [65]. About 30% of transcription factor binding motifs in DNA have either one or, sometimes, two CpG dinucleotides in their DNA sequence [66]. On average, about 12 transcription factors are simultaneously bound to their particular binding motifs on a given highly active enhancer segment of DNA [67]. Altered methylation of the CpG dinucleotide in the binding motif on an enhancer can block or increase the binding of a transcription factor to an enhancer [57]. Blocking or increasing the binding of a transcription factor on an enhancer could block or increase the effectiveness of the enhancer in activating transcription of a gene [68]. Many transcription factors have a preference for binding to DNA motifs that have a methylated CpG, while somewhat fewer have a preference for binding to DNA motifs with CpG dinucleotides without methylation on the CpGs within the motifs [69].

6.6 CpG islands

Early findings evaluating methylation of CpG island methylation with age found that CpG dinucleotides in CpG islands gained methylation with age, while CpG dinucleotides not in CpG islands lost methylation with age [70]. The CpG dinucleotide occurs in many locations of the human genome, but especially in "CpG islands." CpG islands are usually between 500 and 1,500 nucleotides in length [71] and have a frequency of CpG dinucleotides that is about 10-fold higher than in the remainder of the genome [72]. CpG islands occur in 60% of promoters of genes and in about 6% of enhancers [53]. In the 60% of genes with CpG islands in promoters just upstream of their transcription start sites, high methylation of CpGs correlates with low gene expression, and low methylation of CpGs correlates with high gene expression [73].

6.7 Altered methylation accumulates with age as leftovers from DNA repair

In 2005, Fraga et al. [74] showed that identical twins had very closely similar cytosine methylation at particular loci early in life but quite different cytosine methylation at those loci later in life. This indicated that stable changes to cytosine methylation occur in differentiated cells of humans with age. Similarly, between 2011 and 2020, 22 different epigenetic clocks have been proposed, showing that at many genetic loci CpG methylation changes with age [51]. Two types of DNA repair result in alterations to cytosine methylation. Repair of oxidative damage to the guanine in CpG dinucleotides and repair of a double-strand break in DNA can each generate altered methylation of nearby cytosines. These are likely sources of altered CpG methylation in differentiated cells with age. Approximately 100 types of oxidatively caused base lesions and 2-deoxyribose modifications are produced in DNA by reactive oxygen species [75]. Below we describe the repair of some of these types of lesions by different pathways and the resulting demethylation or methylation of cytosine that occur as a side effect.

DOI: http://dx.doi.org/10.5772/intechopen.1012704

7. DNA methylation alterations related to aging

7.1 DNA repair initiated by OGG1 can result in the demethylation of the adjacent cytosine

Guanine is susceptible to oxidative damage within a CpG dinucleotide. This can produce 8-oxo-guanine (or its tautomeric form, 8-OHdG). Repair of this DNA damage is initiated by the DNA repair enzyme oxoguanine glycosylase (OGG1). OGG1 finds an 8-oxodG in DNA within a few seconds [76] and binds there along with cohesin and mediator proteins MED14 and MED17 (Lebraud et al., 2020). When OGG1 is bound to DNA at an oxidized guanine within a methylated CpG site, the OGG1 changes shape, and it recruits the DNA demethylase enzyme TET1 (see **Figure 4**). This allows TET1 to demethylate an adjacent methylated cytosine [62]. However, OGG1 only slowly excises 8-oxodG. OGG1 generally leaves the chromatin in parallel with the removal of 8-oxodG, and 8-oxodG is removed over a period of 11 minutes to 55 minutes [77–79]. While OGG1 mostly leaves DNA in parallel with the removal of 8-oxodG, some level of OGG1 is still seen at the site of damage six hours after the damage occurred [80].

In one example, human mammary epithelial cells were treated with 500 mM H_2O_2 (an oxidizing agent) for 30 minutes [62]. 8-oxodG was increased by about 3.5-fold in the DNA of these cells. This resulted in about 80% demethylation of the 5-methylcytosines in the cellular DNA. Demethylation of CpGs within the promoters

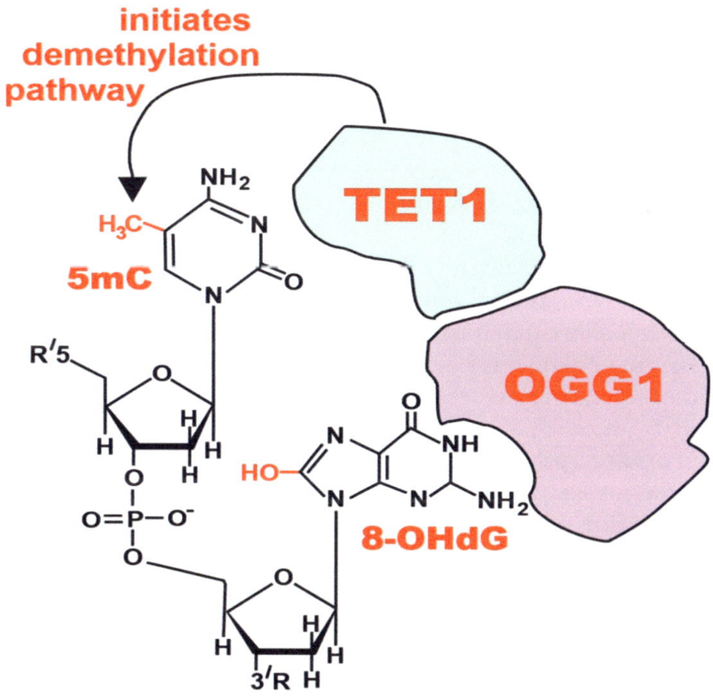

Figure 4.
Initiation of DNA demethylation licensed under the Creative Commons Attribution-Share Alike 4.0 International License Bernstein0275.

of genes by TET enzyme activity increases transcription of the genes into messenger RNA [81]. One particular gene was examined, *beta-secretase 1 (BACE1),* in the cells treated with H2O2 [62]. There was a reduction in the methylation level of the *BACE1* CpG island promoter, and this allowed an increase of about 6.5-fold in the expression of *BACE1* messenger RNA.

There are many sources of reactive oxygen species within cells [82]. These reactive oxygen species are needed for signaling, and they also cause damage. The steady state of oxidative DNA damage reflects the balance between damage and repair. There are at least 100 types of oxidative DNA damage in cells, and 8-oxodG is one of the most frequent, but it accounts for only 5% of the steady state of oxidative damage in DNA [77]. There are 2,400 8-oxodG damages in the steady state of DNA in the nucleus [83]. Some of these likely cause stable demethylation of cytosine at certain locations during aging.

7.2 DNA mismatch repair enzymes repair some oxidative DNA damages and cause new methylations

In a second example, NCCIT cells (derived from a mediastinal mixed germ cell tumor) were treated with 1.0 mM H_2O_2 for 30 minutes. This caused some cell death, but the survivors were examined for oxidative damage to DNA that was repaired by the mismatch repair protein heterodimer MSH2-MSH6. The MSH2-MSH6 heterodimer binds tightly to DNA at the sites of some type(s) of oxidative DNA damage to promote repair of the damage. When MSH2-MSH6 binds to DNA, they recruit DNA methyltransferase 1 (DNMT1) to bind as well. This allows DNMT1 to add a methyl group to a nearby cytosine [84]. This apparently caused increased methylation of cytosines at some CpG islands, causing, in turn, reduced transcription of genes at those locations. Those transcriptional reductions were eliminated with knockdown of MSH6 and/or DNMT1.

7.3 Potassium bromate causes some types of oxidative DNA damage, and base excision repair of those damages alters DNA methylation

Jiang et al. [85] treated human embryonic kidney cells (HEK293H) with an agent causing oxidative DNA damage: potassium bromate (KBrO3). The cells were then allowed to undergo short-patch base excision repair (sp-BER), which utilizes the DNA repair enzyme polymerase beta (POLB). Jiang et al. [85] examined the promoter and early transcribed region of the tumor suppressor breast cancer 1 (*BRCA1*) gene, from base pairs -189 to +27. Zero is the transcription start site, and negative numbers are upstream into the promoter (-1, -2, -3, etc.), while positive numbers are downstream into the early transcribed regions of the gene. They found that POLB, interacting with the DNA methyltransferase protein DNMT3b, was bound to the sites of 8-oxodG created by bromate treatment. They then examined methylation alterations in the region under study. In cells not treated with bromate, CpGs in the *BRCA1* gene located at -189, -134, -29, -19, +16, and +19 had methylated cytosines. Oxidation after bromate treatment resulted in the removal of cytosine methylations at -189, -134, +16, and +19 but also led, after DNA repair was allowed, to the creation of new methylations at CpGs located at -80, -55, -21, and +8.

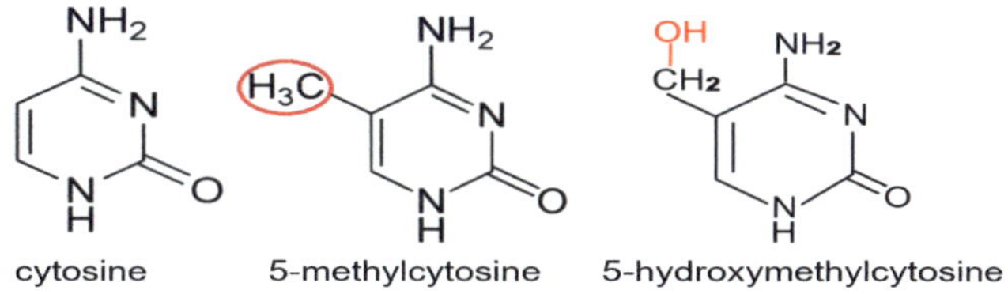

Figure 5.
Cytosine, 5-methylcytosine, 5-hydroxymethyl cytosine.

7.4 Homologous recombinational repair (HRR) of double-strand breaks utilizes DNMT1 and alters cytosine methylation

Four research groups [86–89] each showed that DNMT1 was recruited to the site of a double-strand break in DNA and participated in its repair. Cuozzo et al. [87] demonstrated that the presence of DNMT1 caused half of the cells with homologous recombination repair (HRR) to exhibit high levels of cytosine methylation for 300 base pairs downstream of the repaired cut in DNA. The other half of the cells exhibited very low levels of cytosine methylation. Upon examining the expression of the gene that was broken by the double-strand break, it was found that half of the progeny cells expressed that gene at a high level, while the other half of the progeny cells expressed that gene at a low level. Clones of these cells were maintained for three years, and it was observed that the new methylation and expression patterns were preserved over that time period [90]. Thus, the altered cytosine methylation after HRR was fairly permanent.

7.5 HRR of double-strand breaks utilizes TET2 and TET2 alters nearby cytosine methylations by creating hydroxymethylcytosines

All three TET proteins (TET1, TET2, and TET3) can convert methylcytosine to hydroxymethylcytosine (see **Figure 5**) [91]. Hydroxymethylcytosine is the first step in converting methylcytosine to cytosine. However, hydroxymethylcytosine, which acts like cytosine itself, often remains in DNA as a new epigenetic alteration. TET2 acts at double-strand breaks in DNA to convert methylcytosine to hydroxymethylcytosine [92]. The hydroxymethylcytosine colocalizes with three markers of double-strand break repair: 53BP1, γ H2AX, and Rad51 [92].

While methylation of cytosine may repress the expression of genes, hydroxymethylcytosine often allows increased expression of genes [93]. TET2 acting at DNA double-strand breaks to convert methylcytosine to hydroxymethylcytosine can activate genes adjacent to the double-strand break.

8. DNA double-strand break repair causes epigenetic changes related to aging

There are more than 35 DNA repair enzymes that cooperate to repair DNA double-strand breaks through the accurate process of homologous recombinational repair [94]. The large complex responsible for repairing the break includes the methylating enzyme DNMT1 and the demethylating enzyme TET2 [87, 92]. Yang et al. [14] hypothesized that induced double-strand breaks would cause a relocation of chromatin modifiers, such as the movement of DNMT1 and TET2, to double-strand breaks. This relocation would take the chromatin modifiers away from locations where they were needed for the maintenance of cell identity. They hypothesized that this would cause epigenetic deficiencies elsewhere in the genome. To test this, they introduced an endonuclease, I-Ppol, into the genome of mice, attached to a receptor for tamoxifen. This endonuclease can make DNA double-strand cuts at only 20 locations in the genome (none in protein-coding genes). The endonuclease could be activated by tamoxifen administered through food. They fed tamoxifen in food to 4- to 6-month-old mice that carried the inducible I-*Pp*ol-tamoxifen-receptor for three weeks. They found that the treated mice exhibited more characteristics of aging at 10 months of age (reduced grip strength, reduced motion at night, and at least 10 other modified characteristics) compared to untreated mice. There were no remaining DNA double-strand breaks and no additional mutations. At 74 DNA sites that were typically altered in an epigenetic clock with age, they found altered CpG methylation indicating older age when treated mice were 10 months old. They concluded that DNA double-strand break repair caused epigenetic changes associated with aging at a distance from where the double-strand breaks were occurring.

9. Summary and conclusions

Fundamentally, the aging of an organism appears to be an unavoidable consequence of damage occurring in the genome. We consider aging to be a basic characteristic of life, applying to all organisms. Most aging studies have focused on mammals. However, studies of aging in unicellular eukaryotes and plants have also been reported, and these studies indicate that aging (at least in part due to DNA damage) is universal among eukaryotes and is likely due to either DNA damage directly or indirectly as a result of damage repair leading to epigenetic alterations.

Genome activity includes the basic functions of replication and gene expression. We consider that the processes of replication and gene expression are vulnerable to damage, and therefore the cell's genome is susceptible to injury over time. It is this injury that results in aging. The process of natural selection has led to the evolution of mechanisms for recognizing and repairing DNA injuries, but these processes are not perfect, and damages inevitably accumulate in cells so that, over time, cellular aging cannot be avoided. The evidence presented above indicates that direct damage to DNA is a major cause of cellular aging in mammals. In mammals, the central component of chromosomes is DNA, but other molecules, such as methyl groups that can attach to DNA, control and regulate gene expression. Alteration of these non-DNA molecules, referred to as epigenetic alterations, over time also appears to contribute to aging in mammals. Thus, we conclude that aging, in general, is fundamentally caused both directly by unrepaired DNA damages and also by the repair of DNA damages that leave epigenetic scars.

Author details

Carol Bernstein* and Harris Bernstein
University of Arizona, Tucson, Arizona, United States

*Address all correspondence to: lorac79@yahoo.com

References

[1] Holmes GE, Holmes NR. Accumulation of DNA damages in aging Paramecium tetraurelia. Molecular & General Genetics: MGG. 1986;**204** (1):108–114. DOI:10.1007/BF00330196.

[2] Pongpanich M, Patchsung M, Mutirangura A. Pathologic replication-independent endogenous DNA double-strand breaks repair defect in chronological aging yeast. Frontiers in Genetics. 2018;**9**:501. 10.3389/fgene.2018.00501.

[3] Stewart EJ, Madden R, Paul G, Taddei F. Aging and death in an organism that reproduces by morphologically symmetric division. PLoS Biology. 2005;**3**(2):e45. DOI:10.1371/journal.pbio.0030045.

[4] Chan SW, Dedon PC. The biological and metabolic fates of endogenous DNA damage products. Journal of Nucleic Acids. 2010;**2010**:929047. DOI: 10.4061/2010/929047.

[5] Alexander P. The role of DNA lesions in the processes leading to aging in mice. Symposia of the Society for Experimental Biology. 1967;**21**:29–50.

[6] Gensler HL, Bernstein H. DNA damage as the primary cause of aging. The Quarterly Review of Biology. 1981;**56** (3):279–303. DOI:10.1086/412317.

[7] Bernstein C, Bernstein H. Aging, Sex and DNA Repair (Https://books.google.com/books?id=baxyyuxy71cc&pg=pa3). San Diego: Academic Press; 1991

[8] Ames BN, Gold LS. Endogenous mutagens and the causes of aging and cancer. Mutation Research/Fundamental and Molecular Mechanisms of Mutagenesis. 1991;**250** (1-2):3–16. DOI:10.1016/0027-5107(91)90157-j.

[9] Holmes GE, Bernstein C, Bernstein H. Oxidative and other DNA damages as the basis of aging: A review. Mutation Research/DNAging. 1992;**275**(3-6):305–315. DOI:10.1016/0921-8734(92)90034-m.

[10] Rao KS, Loeb LA. DNA damage and repair in brain: Relationship to aging. Mutation Research/DNAging. 1992;**275**(3-6):317–329. DOI:10.1016/0921-8734(92)90035-n.

[11] Ames BN, Shigenaga MK, Hagen TM. Oxidants, antioxidants, and the degenerative diseases of aging. Proceedings of the National Academy of Sciences of the United States of America. 1993;**90**(17):7915–7922. DOI:10.1073/pnas.90.17.7915.

[12] Niedernhofer LJ, Gurkar AU, Wang Y, Vijg J, Hoeijmakers JHJ, Robbins PD. Nuclear genomic instability and aging. Annual Review of Biochemistry. 2018;**87**:295–322. 10.1146/annurev-biochem-062917-012239.

[13] Vijg J. From DNA damage to mutations: All roads lead to aging. Ageing Research Reviews. 2021;**68**:101316. DOI: 10.1016/j.arr.2021.101316.

[14] Yang JH, Hayano M, Griffin PT, Amorim JA, Bonkowski MS, Apostolides JK, Salfati EL, Blanchette M, Munding EM, Bhakta M, Chew YC, Guo W, Yang X, Maybury-Lewis S, Tian X, Ross JM, Coppotelli G, Meer MV, Rogers-Hammond R, Vera DL, Lu YR, Pippin JW, Creswell ML, Dou Z, Xu C, Mitchell

SJ, Das A, O'Connell BL, Thakur S, Kane AE, Su Q, Mohri Y, Nishimura EK, Schaevitz L, Garg N, Balta AM, Rego MA, Gregory-Ksander M, Jakobs TC, Zhong L, Wakimoto H, El Andari J, Grimm D, Mostoslavsky R, Wagers AJ, Tsubota K, Bonasera SJ, Palmeira CM, Seidman JG, Seidman CE, Wolf NS, Kreiling JA, Sedivy JM, Murphy GF, Green RE, Garcia BA, Berger SL, Oberdoerffer P, Shankland SJ, Gladyshev VN, Ksander BR, Pfenning AR, Rajman LA, Sinclair DA. Loss of epigenetic information as a cause of mammalian aging. Cell. 2023;**186**(2):305–326.e27. DOI:10.1016/j.cell.2022.12.027.

[15] Pan C, Zhou F, Zhang L. The loss of epigenetic information: Not only consequences but a cause of mammalian aging. Signal Transduction and Targeted Therapy. 2023;**8**(1):140. DOI:10.1038/s41392-023-01412-9.

[16] Lu YR, Tian X, Sinclair DA. The information theory of aging. Nature Aging. 2023;**3**(12):1486–1499. DOI:10.1038/s43587-023-00527-6.

[17] Schumacher B, Pothof J, Vijg J, Hoeijmakers JHJ. The central role of DNA damage in the ageing process. Nature. 2021;**592**(7856):695–703. DOI:10.1038/s41586-021-03307-7.

[18] Tubbs A, Nussenzweig A. Endogenous DNA damage as a source of genomic instability in cancer. Cell. 2017;**168**(4):644–656. DOI:10.1016/j.cell.2017.01.002.

[19] Bernstein H, Payne CM, Bernstein C, Garewal H, Dvorak K. 1. Cancer and aging as consequences of un-repaired DNA damage. In Kimura H, Suzuki A, editors. New Research on DNA Damages. Nova Science; 2008. p. 1–47.

[20] Swain U, Subba Rao K. Study of DNA damage via the comet assay and base excision repair activities in rat brain neurons and astrocytes during aging. Mechanisms of Ageing and Development. 2011;**132**(8-9):374–381. DOI:10.1016/j.mad.2011.04.012.

[21] Delint-Ramirez I, Madabhushi R. DNA damage and its links to neuronal aging and degeneration. Neuron. 2025;**113**(1):7–28. DOI:10.1016/j.neuron.2024.12.001.

[22] Savitha S, Panneerselvam C. Mitigation of age-dependent oxidative damage to DNA in rat heart by carnitine and lipoic acid. Mechanisms of Ageing and Development. 2007;**128**(2):206–212. DOI:10.1016/j.mad.2006.11.029.

[23] Sharma N, Coticchio G, Borini A, Tachibana K, Nasmyth KA, Schuh M. Changes in DNA repair compartments and cohesin loss promote DNA damage accumulation in aged oocytes. Current Biology. 2024;**34**(22):5131–5148.e6. DOI:10.1016/j.cub.2024.09.040.

[24] Rossi DJ, Bryder D, Seita J, Nussenzweig A, Hoeijmakers J, Weissman IL. Deficiencies in DNA damage repair limit the function of haematopoietic stem cells with age. Nature. 2007;**447**(7145):725–729. DOI:10.1038/nature05862.

[25] Sharpless NE, DePinho RA. How stem cells age and why this makes us grow old. Nature Reviews Molecular Cell Biology. 2007;**8**(9):703–713. DOI:10.1038/nrm2241.

[26] Helbock HJ, Beckman KB, Shigenaga MK, Walter PB, Woodall AA, Yeo HC, Ames BN. DNA oxidation matters: The HPLC–electrochemical detection assay of 8-oxo-deoxyguanosine and 8-oxo-guanine.

Proceedings of the National Academy of Sciences of the United States of America. 1998;**95**(1):288–293. DOI:10.1073/pnas.95.1.288.

[27] Hashimoto K, Takasaki W, Sato I, Tsuda S. DNA damage measured by comet assay and 8-OH-dG formation related to blood chemical analyses in aged rats. The Journal of Toxicological Sciences. 2007;**32**(3):249–259. DOI:10.2131/jts.32.249.

[28] Oshima J, Sidorova JM, Rj Jr M. Werner syndrome: Clinical features, pathogenesis and potential therapeutic interventions. Ageing Research Reviews. 2017;**33**:105–114. 10.1016/j.arr.2016.03.002.

[29] Gonzalo S, Kreienkamp R. DNA repair defects and genome instability in Hutchinson–Gilford Progeria Syndrome. Current Opinion in Cell Biology. 2015;**34**:75–83. 10.1016/j.ceb.2015.05.007.

[30] D'Errico M, Parlanti E, Teson M, Degan P, Lemma T, Calcagnile A, Iavarone I, Jaruga P, Ropolo M, Pedrini AM, Orioli D, Frosina G, Zambruno G, Dizdaroglu M, Stefanini M, Dogliotti E. The role of CSA in the response to oxidative DNA damage in human cells. Oncogene. 2007;**26**(30):4336–4343. DOI:10.1038/sj.onc.1210232.

[31] Saijo M. The role of Cockayne syndrome group A (CSA) protein in transcription-coupled nucleotide excision repair. Mechanisms of Ageing and Development. 2013;**134**(5-6):196–201. DOI:10.1016/j.mad.2013.03.008.

[32] Burnet FM. Intrinsic mutagenesis: A genetic basis of ageing. Pathology. 1974;**6**(1):1–11. DOI:10.3109/00313027409077150.

[33] Narayanan L, Fritzell JA, Baker SM, Liskay RM, Glazer PM. Elevated levels of mutation in multiple tissues of mice deficient in the DNA mismatch repair gene Pms2. Proceedings of the National Academy of Sciences of the United States of America. 1997;**94**(7):3122–3127. DOI:10.1073/pnas.94.7.3122.

[34] Dollé ME, Giese H, Hopkins CL, Martus HJ, Hausdorff JM, Vijg J. Rapid accumulation of genome rearrangements in liver but not in brain of old mice. Nature Genetics. 1997;**17**(4):431–434. DOI:10.1038/ng1297-431.

[35] Stuart GR, Oda Y, de Boer JG, Glickman BW. Mutation frequency and specificity with age in liver, bladder and brain of lacI transgenic mice. Genetics. 2000;**154**(3):1291–1300. DOI:10.1093/genetics/154.3.1291.

[36] Hill KA, Halangoda A, Heinmoeller PW, Gonzalez K, Chitaphan C, Longmate J, Scaringe WA, Wang JC, Sommer SS. Tissue-specific time courses of spontaneous mutation frequency and deviations in mutation pattern are observed in middle to late adulthood in Big Blue mice. Environmental and Molecular Mutagenesis. 2005;**45**(5):442–454. DOI:10.1002/em.20119.

[37] Vermulst M, Bielas JH, Kujoth GC, Ladiges WC, Rabinovitch PS, Prolla TA, Loeb LA. Mitochondrial point mutations do not limit the natural lifespan of mice. Nature Genetics. 2007;**39**(4):540–543. DOI:10.1038/ng1988.

[38] Dollé ME, Busuttil RA, Garcia AM, Wijnhoven S, van Drunen E, Niedernhofer LJ, van der Horst G, Hoeijmakers JH, van Steeg H, Vijg J. Increased genomic instability is not a prerequisite for shortened lifespan in

DNA repair deficient mice. Mutation Research/Fundamental and Molecular Mechanisms of Mutagenesis. 2006;**596** (1-2):22–35. DOI:10.1016/j.mrfmmm.2005.11.008.

[39] Aufderheide KJ. Clonal aging in Paramecium tetraurelia. II. Evidence of functional changes in the macronucleus with age. Mechanisms of Ageing and Development. 19861987;**37**(3):265–279. DOI:10.1016/0047-6374(86)90044-8.

[40] Smith-Sonneborn J. DNA repair and longevity assurance in Paramecium tetraurelia. Science. 1979;**203** (4385):1115–1117. DOI:10.1126/science.424739.

[41] Gilley D, Blackburn EH. Lack of telomere shortening during senescence in Paramecium. Proceedings of the National Academy of Sciences of the United States of America. 1994;**91** (5):1955–1958. DOI:10.1073/pnas.91.5.1955.

[42] Steinboeck F, Hubmann M, Bogusch A, Dorninger P, Lengheimer T, Heidenreich E. The relevance of oxidative stress and cytotoxic DNA lesions for spontaneous mutagenesis in non-replicating yeast cells. Mutation Research/Fundamental and Molecular Mechanisms of Mutagenesis. 2010;**688** (1-2):47–52. DOI:10.1016/j.mrfmmm.2010.03.006.

[43] Proenca AM, Rang CU, Qiu A, Shi C, Chao L. Cell aging preserves cellular immortality in the presence of lethal levels of damage. PLoS Biology. 2019;**17** (5):e3000266. DOI:10.1371/journal.pbio.3000266.

[44] Waterworth WM, Bray CM, West CE. The importance of safeguarding genome integrity in germination and seed longevity. Journal of Experimental Botany. 2015;**66**(12):3549–3558. DOI:10.1093/jxb/erv080.

[45] Cheah KS, Osborne DJ. DNA lesions occur with loss of viability in embryos of ageing rye seed. Nature. 1978;**272**(5654):593–599. DOI:10.1038/272593a0.

[46] Koppen G, Verschaeve L. The alkaline single-cell gel electrophoresis/comet assay: A way to study DNA repair in radicle cells of germinating Vicia faba. Folia Biologica. 2001;**47**(2):50–54.

[47] Waterworth WM, Masnavi G, Bhardwaj RM, Jiang Q, Bray CM, West CE. A plant DNA ligase is an important determinant of seed longevity. The Plant Journal. 2010;**63**(5):848–860. DOI:10.1111/j.1365-313X.2010.04285.x.

[48] Waterworth WM, Footitt S, Bray CM, Finch-Savage WE, West CE. DNA damage checkpoint kinase ATM regulates germination and maintains genome stability in seeds. Proceedings of the National Academy of Sciences of the United States of America. 2016;**113** (34):9647–9652. DOI:10.1073/pnas.1608829113.

[49] Aoyagi Blue Y, Kusumi J, Satake A. Copy number analyses of DNA repair genes reveal the role of poly(ADP-ribose) polymerase (PARP) in tree longevity. iScience. 2021;**24**(7):102779. DOI:10.1016/j.isci.2021.102779.

[50] Al Aboud NM, Tupper C, and Genetics JI, Epigenetic mechanism. [Updated 2023 Aug 14]. In: StatPearls [Internet]. Treasure Island (FL): StatPearls Publishing; 2025. Available from: https://www.ncbi.nlm.nih.gov/bo

oks/NBK532999/. [Accessed: 2025-August-5]

[51] Warner B, Ratner E, Datta A, Lendasse A. A systematic review of phenotypic and epigenetic clocks used for aging and mortality quantification in humans. Aging (Albany. NY). 2024;**16**(17):12414–12427. DOI:10.18632/aging.206098.

[52] Jain N, Li JL, Tong L, Jasmine F, Kibriya MG, Demanelis K, Oliva M, Chen LS, Pierce BL. DNA methylation correlates of chronological age in diverse human tissue types. Epigenetics and Chromatin. 2024;**17**(1):25. DOI:10.1186/s13072-024-00546-6.

[53] Steinhaus R, Gonzalez T, Seelow D, Robinson PN. Pervasive and CpG-dependent promoter-like characteristics of transcribed enhancers. Nucleic Acids Research. 2020;**48**(10):5306–5317. DOI:10.1093/nar/gkaa223.

[54] Poss ZC, Ebmeier CC, Taatjes DJ. The mediator complex and transcription regulation. Critical Reviews in Biochemistry and Molecular Biology. 2013;**48**(6):575–608. DOI:10.3109/10409238.2013.840259.

[55] Haberle V, Stark A. Eukaryotic core promoters and the functional basis of transcription initiation. Nature Reviews Molecular Cell Biology. 2018;**19**(10):621–637. DOI:10.1038/s41580-018-0028-8.

[56] Jang HS, Shin WJ, Lee JE, Do JT. CpG and non-CpG methylation in epigenetic gene regulation and brain function. Genes (Basel). 2017;**8**(6):148. DOI:10.3390/genes8060148.

[57] Yin Y, Morgunova E, Jolma A, Kaasinen E, Sahu B, Khund-Sayeed S, Das PK, Kivioja T, Dave K, Zhong F, Nitta KR, Taipale M, Popov A, Ginno PA, Domcke S, Yan J, Schübeler D, Vinson C, Taipale J. Impact of cytosine methylation on DNA binding specificities of human transcription factors. Science. 2017;**356**(6337):eaaj2239. DOI:10.1126/science.aaj2239.

[58] Zheng Z, Li J, Liu T, Fan Y, Zhai QC, Xiong M, Wang QR, Sun X, Zheng QW, Che S, Jiang B, Zheng Q, Wang C, Liu L, Ping J, Wang S, Gao DD, Ye J, Yang K, Zuo Y, Ma S, Yang YG, Qu J, Zhang F, Jia P, Liu GH, Zhang W. DNA methylation clocks for estimating biological age in Chinese cohorts. Protein and Cell. 2024;**15**(8):575–593. DOI:10.1093/procel/pwae011.

[59] Santos F, Hendrich B, Reik W, Dean W. Dynamic reprogramming of DNA methylation in the early mouse embryo. Developmental Biology. 2002;**241**(1):172–182. DOI:10.1006/dbio.2001.0501.

[60] Lister R, Pelizzola M, Dowen RH, Hawkins RD, Hon G, Tonti-Filippini J, Nery JR, Lee L, Ye Z, Ngo QM, Edsall L, Antosiewicz-Bourget J, Stewart R, Ruotti V, Millar AH, Thomson JA, Ren B, Ecker JR. Human DNA methylomes at base resolution show widespread epigenomic differences. Nature. 2009;**462**(7271):315–322. DOI:10.1038/nature08514.

[61] Ming X, Matter B, Song M, Veliath E, Shanley R, Jones R, Tretyakova N. Mapping structurally defined guanine oxidation products along DNA duplexes: Influence of local sequence context and endogenous cytosine methylation. Journal of the American Chemical Society. 2014;**136**(11):4223–4235. DOI:10.1021/ja411636j.

[62] Zhou X, Zhuang Z, Wang W, He L, Wu H, Cao Y, Pan F, Zhao J, Hu Z,

Sekhar C, Guo Z. OGG1 is essential in oxidative stress induced DNA demethylation. Cellular Signalling. 2016;**28**(9):1163–1171. DOI:10.1016/j.cellsig.2016.05.021.

[63] Li Y, Tian X, Luo J, Bao T, Wang S, Wu X. Molecular mechanisms of aging and anti-aging strategies. Cell Communication and Signaling : CCS. 2024;**22**(1):285. DOI:10.1186/s12964-024-01663-1.

[64] Amaral P, Carbonell-Sala S, FM DLV, Faial T, Frankish A, Gingeras T, Guigo R, Harrow JL, Hatzigeorgiou AG, Johnson R, Murphy TD, Pertea M, Pruitt KD, Pujar S, Takahashi H, Ulitsky I, Varabyou A, Wells CA, Yandell M, Carninci P, Salzberg SL. The status of the human gene catalogue. Nature. 2023;**622**(7981):41–47. DOI:10.1038/s41586-023-06490-x.

[65] Pratt HE, Andrews GR, Phalke N, Purcaro MJ, van der Velde A, Moore JE, Weng Z. Factorbook: An updated catalog of transcription factor motifs and candidate regulatory motif sites. Nucleic Acids Research. 2022;**50**(D1): D141–D149. DOI:10.1093/nar/gkab1039.

[66] Héberlé É, Bardet AF. Sensitivity of transcription factors to DNA methylation. Essays in Biochemistry. 2019;**63**(6):727–741. DOI:10.1042/EBC20190033.

[67] Singh G, Mullany S, Moorthy SD, Zhang R, Mehdi T, Tian R, Duncan AG, Moses AM, Mitchell JA. A flexible repertoire of transcription factor binding sites and a diversity threshold determines enhancer activity in embryonic stem cells. Genome Research. 2021;**31**(4):564–575. DOI:10.1101/gr.272468.120.

[68] He A, Kong SW, Ma Q, Pu WT. Co-occupancy by multiple cardiac transcription factors identifies transcriptional enhancers active in heart. Proceedings of the National Academy of Sciences of the United States of America. 2011;**108**(14):5632–5637. DOI:10.1073/pnas.1016959108.

[69] Li Y, Zou Q, Dai Q, Stalin A, Luo X. Identifying the DNA methylation preference of transcription factors using ProtBERT and SVM. PLOS Computational Biology. 2025;**21**(5): e1012513. DOI:10.1371/journal.pcbi.1012513.

[70] Christensen BC, Houseman EA, Marsit CJ, Zheng S, Wrensch MR, Wiemels JL, Nelson HH, Karagas MR, Padbury JF, Bueno R, Sugarbaker DJ, Yeh RF, Wiencke JK, Kelsey KT. Aging and environmental exposures alter tissue-specific DNA methylation dependent upon CpG island context. PLoS Genetics. 2009;**5**(8):e1000602. DOI:10.1371/journal.pgen.1000602.

[71] Elango N, Yi SV. Functional relevance of CpG island length for regulation of gene expression. Genetics. 2011;**187**(4):1077–1083. DOI:10.1534/genetics.110.126094.

[72] Antequera F, Bird A. Cpg islands as genomic footprints of promoters that are associated with replication origins. Current Biology. 1999;**9**(17): R661–7. DOI:10.1016/s0960-9822(99)80418-7.

[73] Ando M, Saito Y, Xu G, Bui NQ, Medetgul-Ernar K, Pu M, Fisch K, Ren S, Sakai A, Fukusumi T, Liu C, Haft S, Pang J, Mark A, Gaykalova DA, Guo T, Favorov AV, Yegnasubramanian S, Fertig EJ, Ha P, Tamayo P, Yamasoba T, Ideker T, Messer K, Califano JA.

Chromatin dysregulation and DNA methylation at transcription start sites associated with transcriptional repression in cancers. Nature Communications. 2019;**10**(1):2188. DOI:10.1038/s41467-019-09937-w.

[74] Fraga MF, Ballestar E, Paz MF, Ropero S, Setien F, Ballestar ML, Heine-Suñer D, Cigudosa JC, Urioste M, Benitez J, Boix-Chornet M, Sanchez-Aguilera A, Ling C, Carlsson E, Poulsen P, Vaag A, Stephan Z, Spector TD, Wu YZ, Plass C, Esteller M. Epigenetic differences arise during the lifetime of monozygotic twins. Proceedings of the National Academy of Sciences of the United States of America. 2005;**102** (30):10604–10609. DOI:10.1073/pnas.0500398102.

[75] Cadet J, Wagner JR. DNA base damage by reactive oxygen species, oxidizing agents, and UV radiation. Cold Spring Harbor Perspectives in Biology. 2013;**5**(2):a012559. DOI:10.1101/cshperspect.a012559.

[76] D'Augustin O, Huet S, Campalans A, Radicella JP. Lost in the crowd: How does human 8-oxoguanine DNA glycosylase 1 (OGG1) find 8-oxoguanine in the genome? International Journal of Molecular Sciences. 2020;**21**(21):8360. DOI:10.3390/ijms21218360.

[77] Hamilton ML, Guo Z, Fuller CD, Van Remmen H, Ward WF, Austad SN, Troyer DA, Thompson I, Richardson A. A reliable assessment of 8-oxo-2-deoxyguanosine levels in nuclear and mitochondrial DNA using the sodium iodide method to isolate DNA. Nucleic Acids Research. 2001;**29**(10):2117–2126. DOI:10.1093/nar/29.10.2117.

[78] Lan L, Nakajima S, Oohata Y, Takao M, Okano S, Masutani M, Wilson SH, Yasui A. In situ analysis of repair processes for oxidative DNA damage in mammalian cells. Proceedings of the National Academy of Sciences of the United States of America. 2004;**101** (38):13738–13743. DOI:10.1073/pnas.0406048101.

[79] Jaruga P, Dizdaroglu M. Repair of products of oxidative DNA base damage in human cells. Nucleic Acids Research. 1996;**24**(8):1389–1394. DOI:10.1093/nar/24.8.1389.

[80] Lebraud E, Pinna G, Siberchicot C, Depagne J, Busso D, Fantini D, Irbah L, Robeska E, Kratassiouk G, Ravanat JL, Epe B, Radicella JP, Campalans A. Chromatin recruitment of OGG1 requires cohesin and mediator and is essential for efficient 8-oxoG removal. Nucleic Acids Research. 2020;**48** (16):9082–9097. DOI:10.1093/nar/gkaa611.

[81] Maeder ML, Angstman JF, Richardson ME, Linder SJ, Cascio VM, Tsai SQ, Ho QH, Sander JD, Reyon D, Bernstein BE, Costello JF, Wilkinson MF, Joung JK. Targeted DNA demethylation and activation of endogenous genes using programmable TALE-TET1 fusion proteins. Nature Biotechnology. 2013;**31**(12):1137–1142. DOI:10.1038/nbt.2726.

[82] Sies H, Berndt C, Jones DP. Oxidative stress. Annual Review of Biochemistry. 2017;**86**:715–748. 10.1146/annurev-biochem-061516-045037.

[83] Swenberg JA, Lu K, Moeller BC, Gao L, Upton PB, Nakamura J, Starr TB. Endogenous versus exogenous DNA adducts: Their role in carcinogenesis, epidemiology, and risk assessment. Toxicological Sciences. 2011; Suppl **120** (Suppl Supplement 1):S130–45. DOI:10.1093/toxsci/kfq371.

[84] Ding N, Bonham EM, Hannon BE, Amick TR, Baylin SB, O'Hagan HM. Mismatch repair proteins recruit DNA methyltransferase 1 to sites of oxidative DNA damage. Journal of Molecular Cell Biology. 2016;**8**(3):244–254. DOI:10.1093/jmcb/mjv050.

[85] Jiang Z, Lai Y, Beaver JM, Tsegay PS, Zhao ML, Horton JK, Zamora M, Rein HL, Miralles F, Shaver M, Hutcheson JD, Agoulnik I, Wilson SH, Liu Y. Oxidative DNA damage modulates DNA methylation pattern in human breast cancer 1 (BRCA1) gene via the crosstalk between DNA polymerase β and a de novo DNA methyltransferase. Cells. 2020;**9**(1):225. DOI:10.3390/cells9010225.

[86] Mortusewicz O, Schermelleh L, Walter J, Cardoso MC, Leonhardt H. Recruitment of DNA methyltransferase I to DNA repair sites. Proceedings of the National Academy of Sciences of the United States of America. 2005;**102** (25):8905–8909. DOI:10.1073/ pnas.0501034102.

[87] Cuozzo C, Porcellini A, Angrisano T, Morano A, Lee B, Di Pardo A, Messina S, Iuliano R, Fusco A, Santillo MR, Muller MT, Chiariotti L, Gottesman ME, and Avvedimento EV. DNA damage, homology-directed repair, and DNA methylation. PLoS Genetics. 2007;**3**(7):e110. Erratum in: PLoS Genet. 201710;13(2):e1006605. doi: 10.1371/journal.pgen.1006605. PMID: 17616978; PMCID: PMC1913100. DOI:10.1371/journal. pgen.0030110.

[88] O'Hagan HM, Mohammad HP, Baylin SB. Double strand breaks can initiate gene silencing and SIRT1-dependent onset of DNA methylation in an exogenous promoter CpG island. PLoS Genetics. 2008;**4**(8):e1000155. DOI:10.1371/journal.pgen.1000155.

[89] Ha K, Lee GE, Palii SS, Brown KD, Takeda Y, Liu K, Bhalla KN, Robertson KD. Rapid and transient recruitment of DNMT1 to DNA double-strand breaks is mediated by its interaction with multiple components of the DNA damage response machinery. Human Molecular Genetics. 2011;**20**(1):126–140. DOI:10.1093/hmg/ddq451.

[90] Russo G, Landi R, Pezone A, Morano A, Zuchegna C, Romano A, Muller MT, Gottesman ME, Porcellini A, Avvedimento EV. DNA damage and repair modify DNA methylation and chromatin domain of the targeted locus: Mechanism of allele methylation polymorphism. Scientific Reports. 2016;**6**:33222. DOI: 10.1038/ srep33222.

[91] Putiri EL, Tiedemann RL, Thompson JJ, Liu C, Ho T, Choi JH, Robertson KD. Distinct and overlapping control of 5-methylcytosine and 5-hydroxymethylcytosine by the TET proteins in human cancer cells. Genome Biology. 2014;**15**(6):R81. DOI:10.1186/g b-2014-15-6-r81.

[92] Kafer GR, Li X, Horii T, Suetake I, Tajima S, Hatada I, Carlton PM. 5-hydroxymethylcytosine marks sites of DNA damage and promotes genome stability. Cell Reports. 2016;**14** (6):1283–1292. DOI:10.1016/j. celrep.2016.01.035.

[93] Mellén M, Ayata P, Heintz N. 5-hydroxymethylcytosine accumulation in postmitotic neurons results in functional demethylation of expressed genes. Proceedings of the National

Academy of Sciences of the United States of America. 2017;**114** (37):E7812–E7821. DOI:10.1073/pnas.1708044114.

[94] Human DNA Repair Genes [Internet]. 2020. Available from: https://www.mdanderson.org/documents/Labs/Wood-Laboratory/human-dna-repair-genes.html. [Accessed: 2025-July-26]

Chapter 4

The Art of Imperfect Joining: NHEJ in DNA Repair and Disease

Yan-Ji Lu, Xian Hong and Tao Wang

Abstract

Nonhomologous end joining (NHEJ), the primary mammalian DNA double-strand break repair mechanism, serves as a crucial guardian of genomic integrity while simultaneously posing mutagenic risks due to its inherent error-proneness. This chapter delineates the core molecular architecture of NHEJ, involving sequential end recognition, processing, and ligation, modulated by specialized accessory factors that regulate repair efficiency, and reviews the pivotal 53BP1-BRCA1 regulatory axis that determines repair pathway choice between NHEJ and homologous recombination. This chapter also outlines NHEJ dysfunction in diverse pathological contexts, including tumors, immunodeficiency disorders, and neurodegenerative conditions, while exploring cutting-edge therapeutic interventions such as DNA-dependent protein kinase catalytic subunit inhibition and synthetic lethality strategies. Collectively, this chapter establishes NHEJ as both a fundamental genome preservation system and a potential therapeutic target, with profound implications for precision medicine approaches to DNA repair-related diseases.

Keywords: non-homologous end joining, DNA damage, double-strand breaks, genome stability, disease

1. Introduction

As the primary carrier of genetic information, chromatin must maintain its DNA integrity, which is crucial for cell survival. However, this integrity is constantly challenged by both exogenous and endogenous damaging agents. External stresses include ionizing radiation, ultraviolet irradiation, and chemical mutagens. Endogenously, DNA damage arises from replication fork collapse, spontaneous hydrolytic depurination/depyrimidination, and aberrant recombination events [1]. These etiologies induce distinct damage profiles; for example, ionizing radiation predominantly causes DNA double-strand breaks (DSBs), ultraviolet light triggers pyrimidine dimers and other photoproducts, while chemical mutagens lead to base mismatches, alkylations, DSBs, DNA single-strand breaks (SSBs), or cross-linking. Among these lesions, DSBs represent the most deleterious form, potentially instigating cell apoptosis or oncogenic transformation [2].

DSBs represent the most severe threat to genomic integrity due to their fundamental disruption of the DNA double helix. Unlike SSBs or base lesions, which leave an intact complementary strand to serve as a template for accurate repair, DSBs completely sever both strands of the DNA molecule. This physical separation creates two free DNA ends that lack intrinsic sequence context for precise realignment. Consequently, cells face immense challenges in faithfully restoring the original sequence during repair. Failure to correctly rejoin broken ends or misrepair involving different chromosomes can lead to large-scale chromosomal aberrations, including translocations, deletions, inversions, or acentric fragments. These gross chromosomal rearrangements are potent drivers of genomic instability, a hallmark of cancer, and can directly disrupt essential genes or regulatory elements. Unrepaired or misrepaired DSBs are highly cytotoxic, acting as a potent trigger for programmed cell death (apoptosis) or inducing permanent cell cycle arrest (senescence). Thus, the unique ability of DSBs to cause irreversible genetic alterations, genomic chaos, and cell death underscores their status as the most deleterious form of DNA damage.

To counteract the threat of DSBs, cells have evolved two primary repair pathways: nonhomologous end joining (NHEJ), a rapid but error-prone mechanism, and homologous recombination (HR), a high-fidelity yet slower process. The NHEJ concept was first proposed by Moore and Haber in 1996 as the primary DSB repair mechanism in mammalian cells [3, 4]. Unlike the HR repair pathway, NHEJ is characterized by its nonhomologous nature – repair proceeds without homologous sequence guidance (**Figure 1**). With relatively rapid repair kinetics, NHEJ markedly reduces the risk of unrepaired DSBs inducing cell apoptosis, thereby playing an irreplaceable role in maintaining cellular physiological functions.

Despite the protective function of NHEJ in ensuring cell survival in response to DSBs, the high efficiency of NHEJ is attained at the cost of repair fidelity. On one

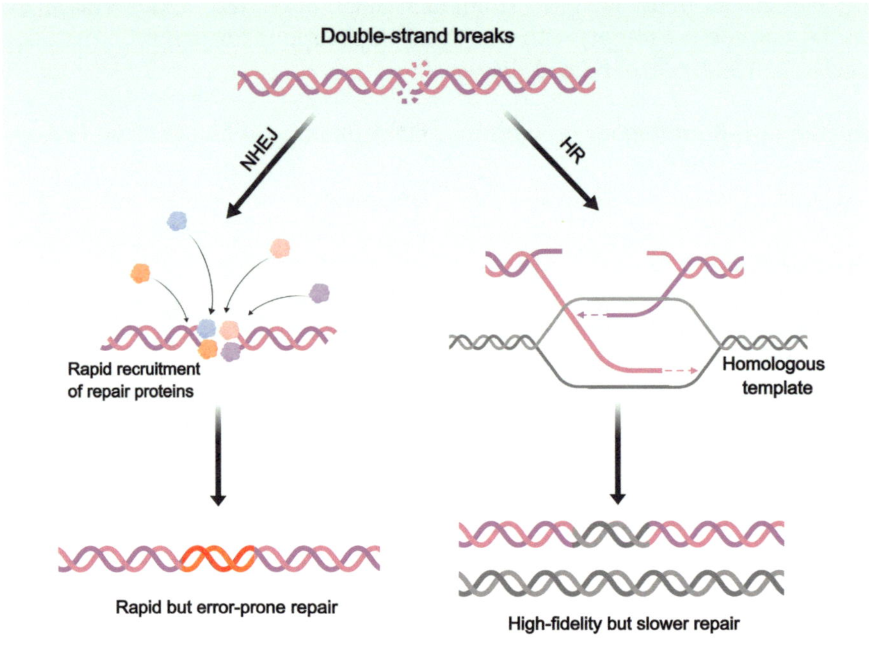

Figure 1.
Overall distinctions between NHEJ and HR.

hand, the diversity of DNA end configurations and the frequent end-processing steps of NHEJ result in repair products that rarely restore predamage sequences [5]. On the other hand, NHEJ directly ligates broken DNA ends without homologous templates, a rapid-response advantage that often incurs nucleotide insertions or deletions, leading to frameshift mutations or gene functional loss.

Despite its trade-offs in accuracy and efficiency, NHEJ remains the dominant DSB repair pathway, responsible for nearly all DSB repair outside the S and G2 phases and approximately 80% of DSB repair during S and G2 [6, 7]. In this chapter, we provide an overview of the current understanding of NHEJ, including its molecular mechanisms, regulatory networks, associated diseases, and therapeutic applications.

2. How does NHEJ work?

The NHEJ pathway orchestrates DSB repair through three coordinated phases: (1) break recognition, (2) end processing, and (3) ligation. Initially, the Ku70/Ku80 heterodimer (Ku protein) recognizes and binds to the broken DNA ends, serving as the primary damage sensor. Subsequently, Ku recruits DNA-dependent protein kinase catalytic subunit (DNA-PKcs) to form the DNA-PK complex, which stabilizes the broken ends. Depending on the complexity of the DNA termini, additional processing enzymes such as Artemis, Pol μ, and Pol λ may be recruited to modify the ends, often resulting in nucleotide loss or addition-a key reason why NHEJ frequently fails to restore the original DNA sequence. Finally, the XRCC4-LIG4-XLF complex is recruited to catalyze the ligation of processed ends, completing the repair (**Figure 2**) [8]. Building upon this general framework, the following sections will dissect the molecular intricacies of each NHEJ step.

2.1. DNA DSBs recognition by Ku protein: The initiation of NHEJ

In eukaryotic cells, the Ku heterodimer (composed of Ku70/XRCC6 and Ku80/XRCC5 subunits) serves as the primary sensor for DNA DSBs. This ring-shaped protein complex exhibits remarkable binding kinetics, recognizing and stabilizing broken DNA ends within one second of damage occurrence – a response time much earlier than that of most other possible sensors [9]. The resulting Ku-DNA complex maintains broken end proximity to prevent deleterious translocation events [10]. The DNA-binding properties of Ku were first characterized in seminal work by Mimori and Hardin (1986), who demonstrated both its exceptionally high affinity for linear DNA (40 fmol binding >90% of double-stranded DNA under saturating conditions) and its strict dependence on DNA ends for binding (showing no detectable interaction with intact circular plasmids) [11]. Upon DNA engagement, Ku undergoes significant conformational changes, as evidenced by its failure to form stable complexes with DNA-PKcs in the absence of DNA ends [12], and similar requirements for interactions with polymerases μ and λ, as well as XRCC4-DNA ligase IV [13, 14]. So as a result of these structural changes, Ku is converted into a molecular scaffold for NHEJ factor assembly [15]. The rapid detection and structural stabilization provided by Ku binding prevent deleterious end translocation while creating a platform for subsequent repair steps. This dual function, combining

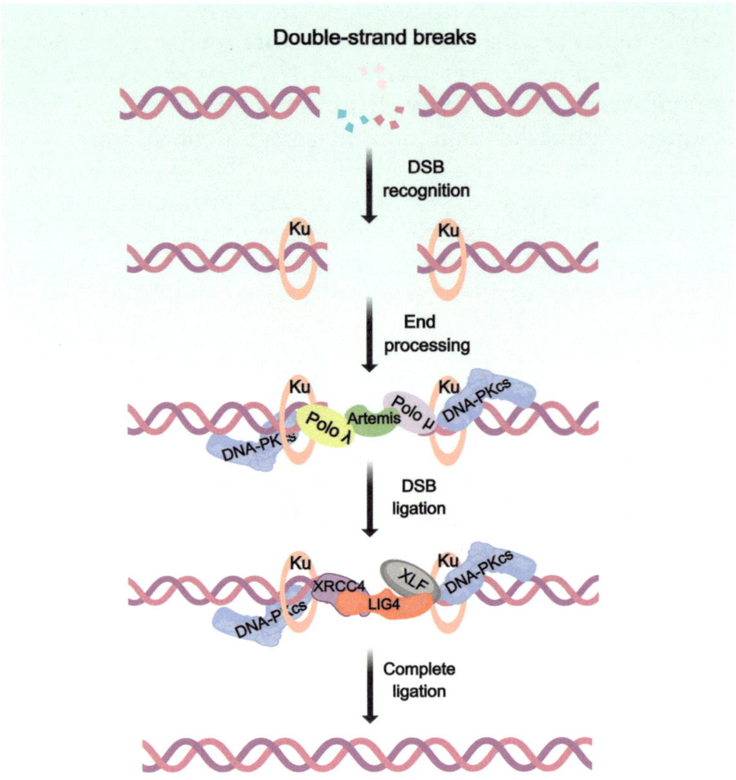

Figure 2.
Schematic of the core NHEJ process.

instantaneous damage sensing with sophisticated protein recruitment, establishes Ku as the critical first responder in eukaryotic DSB repair.

DNA-PKcs, a crucial member of the phosphatidylinositol 3-kinase-related kinase (PIKK) family [16], plays a pivotal role in NHEJ alongside other DNA damage response proteins, ATM and ATR. As the primary response factor following Ku binding to damaged DNA, DNA-PKcs exhibits dynamic and flexible interactions with the Ku complex. Structural studies reveal that the C-terminal 12 amino acids of Ku80 are necessary and sufficient for DNA-PKcs binding [17, 18], though this interaction remains weak without Ku's prior engagement with DNA ends. Remarkably, Ku's presence on DNA enhances DNA-PKcs's affinity for DNA termini by 100-fold [19]. While DNA-PKcs maintains an unphosphorylated, inactive state in isolation, its association with Ku triggers rapid activation [20, 21], initiating the NHEJ repair cascade.

2.2. DNA end processing by Artemis and polymerases

The broken ends of DNA often present complex structural challenges, including mismatches and covalent modifications that must be resolved prior to repair. A remarkable feature of NHEJ is its ability to process and ligate an exceptionally diverse array of damaged DNA substrates. To accomplish this, specialized enzymatic machinery must be recruited to modify and prepare these heterogeneous DNA termini for subsequent joining.

Artemis serves as the primary nuclease for resolving complex DNA termini during NHEJ. Upon Ku-mediated recruitment of DNA-PKcs to DSBs, the activated DNA-PKcs-Artemis complex targets diverse damaged ends, including overhangs, gaps, and chemically modified structures [22, 23]. DNA-PKcs autophosphorylation triggers Artemis activation by relieving autoinhibition of its catalytic domain, enabling cutting at single-to-double-strand DNA boundaries. This processing generates ligation-competent ends while preventing extensive resection, which occurs to initiate HR pathways, through Ku's remarkable binding with DNA ends. Notably, Artemis also interacts with DNA ligase IV (LIG4) via its C-terminal region, coordinating incision with subsequent ligation steps.

Template-dependent and template-independent gap filling is mediated by Pol μ and Pol λ, the primary Pol X family polymerases in NHEJ [24, 25]. Both enzymes dock onto Ku-bound DNA ends via N-terminal BRCT domains, positioning them at repair junctions [26]. They exhibit unique enzymatic flexibility: incorporating both dNTPs and rNTPs (with erroneous ribonucleotides later corrected by base excision repair) in a template-dependent or template-independent manner [27]. This versatility permits nucleotide addition across diverse end configurations, including noncomplementary overhangs where conventional polymerases would stall.

The iterative, context-dependent actions of Artemis, Pol μ, and Pol λ collectively explain NHEJ's ability to join structurally diverse breaks, albeit with inherent mutagenic consequences. Artemis processing creates transient microhomology or blunt ends, while Pol μ/λ fill gaps or extend termini through templated or nontemplated synthesis. This sequential yet flexible processing enables ligation of otherwise incompatible ends but introduces insertions, deletions, or sequence alterations at repair junctions, underscoring NHEJ's error-prone nature.

2.3. Ligation of damaged DNA ends

After the DSB ends undergo a series of processing steps, the core enzyme of the NHEJ pathway, LIG4, is gradually recruited and forms the XRCC4-XLF-LIG4 complex to facilitate efficient ligation of the damaged DNA [28]. Specifically, XRCC4 and XLF assemble into an omega (ω)-shaped flexible scaffold, with the lateral arms formed by XRCC4 coiled-coil extensions and the base supported by a complex of XLF and XRCC4. The BRCT (BRCA1 C Terminus) domain of LIG4 docks onto the ω -shaped scaffold and is suspended above it, forming a presynaptic complex along with the previously recruited Ku dimer and DNA-PKcs. Subsequently, autophosphorylation of DNA-PKcs induces a conformational change, leading to its dissociation from the presynaptic complex, leaving the remaining components to form the synaptic complex [29]. Upon binding ATP, the catalytic domain of LIG4 transfers AMP to the 5'-phosphate group at the DNA end, generating a DNA-AMP intermediate. With the assistance of XRCC4, the 3'-hydroxyl group at the end of LIG4 catalyzes the cleavage of the 5'-phosphate group of the DNA-AMP intermediate to form a phosphodiester bond, and at the same time releases AMP to complete the double-strand linkage [30].

3. Regulation of NHEJ

3.1. Emerging accessory factors fine-tune the NHEJ machinery

Beyond the core NHEJ machinery (Ku70/80, DNA-PKcs, Artemis, Pol λ/μ, LIG4-XRCC4-XLF), recent studies reveal critical regulatory roles for novel accessory factors [31]. The paralog of XRCC4 and XLF (PAXX) stabilizes the NHEJ complex at DNA breaks by interacting with Ku, enhancing ligation efficiency [32, 33]. While not essential alone, PAXX plays an accessory role during NHEJ that is largely overlapped by XLF's function in maintaining complex integrity [34, 35]. MRI/CYREN orchestrates cell-cycle-dependent pathway choice: In the G1 phase, it promotes NHEJ by stabilizing repair protein recruitment to chromatin via its Ku-binding motif [36]; in the S/G2 phases, it sequesters Ku away from breaks, inhibiting NHEJ and favoring HR [37]. The RNA/DNA-binding protein TDP-43 facilitates NHEJ in neuronal cells by promoting nuclear import of the LIG4-XRCC4 complex and stimulating its ligation activity [38–40]. Pathogenic TDP-43 mutation Q331K causes cytoplasmic aggregation, mislocalizing LIG4-XRCC4 and impairing DSB repair – a mechanism linked to amyotrophic lateral sclerosis (ALS) [38]. RNase H2 resolves a unique vulnerability: Pol μ's incorporation of rNTPs during gap filling (due to reduced sugar discrimination). RNase H2 excises these embedded rNMPs postligation, completing repair and preventing genomic instability [41]. IFFO1 tethers broken DNA ends to the nuclear lamina (via XRCC4 and lamin A/C), preventing aberrant chromosomal translocations by immobilizing ends during repair [42]. The chromatin remodeler ERCC6L2 facilitates NHEJ and class switch recombination, with deficiency causing DSB sensitivity and synthetic lethality in XLF-deficient contexts [43–45]. Forkhead box protein L2 (FOXL2) inhibits NHEJ by disrupting Ku heterodimer formation through interaction with Ku protein [46]. Cyclic GMP-AMP synthase (cGAS), the cytosolic DNA sensor, promotes NHEJ by stabilizing DNA-PKcs through Ku80 interactions and USP7-mediated deubiquitination [47]. The mitotic deacetylase complex (MiDAC) maintains NHEJ efficiency by preventing H2AK5K9 hyperacetylation-induced liquid–liquid phase separation of Ku/BRD4, thereby ensuring proper LIG4-XRCC4-XLF complex recruitment to DNA breaks [48]. ZNF384 mediates Ku70/80 loading and NHEJ factor assembly via poly(adenosine diphosphate (ADP)-ribose) polymerase 1 (PARP1)-dependent DSB recruitment and direct Ku interaction, with deficiency causing radiation sensitivity and telomere defects [49]. The acetyltransferase ESCO2 promotes NHEJ by acetylating structural maintenance of chromosomes 3 (SMC3) to stabilize the cohesin complex and regulate chromatin structure to promote 53BP1 microdomain formation [50].

Wiskott–Aldrich syndrome protein and SCAR homolog (WASH) functions as an actin nucleation-promoting factor that regulates retromer-dependent endosomal sorting and trafficking by generating branched filamentous actin (F-actin) on endosomes [51, 52]. Our studies demonstrate that WASH and its obligate partner FAM21 interact with Ku70/Ku80 and DNA-PKcs, with Ku70/Ku80 mediating their recruitment to DSB sites [53, 54]. Consequently, WASH facilitates the DNA damage response by promoting H2AX and KAP1 phosphorylation, enhancing chromatin relaxation, and mediating the recruitment of NHEJ repair factors XLF and LIG4 to DSBs, ultimately increasing NHEJ repair efficiency [53]. While vacuolar protein sorting-associated protein 35 (VPS35) interacts with FAM21, our investigations, albeit confirming a functional role for VPS35 in DNA repair processes, reveal a functional independence from the FAM21/WASH complex. VPS35 functions as a negative regulator of NHEJ by binding and sequestering the Ku70/Ku80 heterodimer away from DNA damage sites. This interaction prevents DNA-PKcs activation and subsequent recruitment of downstream repair factors, including XLF and LIG4, ultimately suppressing NHEJ efficiency [55]. Notably, upon DNA damage, both VPS35 and FOXL2 dissociate from Ku proteins through distinct mechanisms: VPS35 relocates to the cytoplasm via nuclear export, while FOXL2 undergoes SIRT1-mediated deacetylation. Both these events release Ku proteins for proper recruitment to DSBs, thereby restoring NHEJ activity [46, 55].

These discoveries redefine NHEJ as a dynamically regulated pathway, where accessory factors confer context-specific control (**Table 1**). Their dysfunction underlies diseases ranging from immunodeficiencies to neurodegeneration and cancer, revealing new layers of therapeutic vulnerability.

3.2. NHEJ and HR pathway choice

In normal cells, most DSBs are repaired either by the NHEJ or by the HR pathway, and these two mechanisms are in evident competition during the S/G2 phases of the

Proteins	Function	References
PAXX	Stabilize the NHEJ complex at DNA breaks by interacting with Ku to improve ligation efficiency.	[32, 33]
MRI/ CYREN	Stabilize repair protein to chromatin in the G1 phase and promote NHEJ; isolation of Ku in the S/G2 phase inhibits NHEJ and promotes HR.	[36, 37]
TDP-43	Promoting LIG4-XRCC4 complex nuclear import and stimulating its ligation activity.	[38–40]
RNase H2	Resection of embedded rNMPs after ligation completes repair and maintains genome stability.	[41]
IFFO1	Anchoring broken DNA to the nuclear lamina via XRCC4 and lamin A/C prevents chromosomal translocation during repair.	[42]
FOXL2	Inhibition of NHEJ by disrupting Ku heterodimer formation through interaction with Ku.	[46]
cGAS	Promoting NHEJ by interacting with Ku80 to enhance the interaction between DNA-PKcs and the deubiquitinase USP7.	[47]
MiDAC	Preventing Ku/BRD4 phase separation and ensuring the recruitment of the LIG4-XRCC4-XLF complex to DNA breaks.	[48]
ZNF384	Mediating Ku70/80 loading and NHEJ factor assembly via PARP1-dependent recruitment and direct Ku interaction.	[49]
ESCO2	Promoting NHEJ by acetylating SMC3 to stabilize the cohesin complex and regulating chromatin structure to promote 53BP1 microdomain formation.	[50]
WASH	Promotes H2AX/KAP1 phosphorylation, enhances chromatin relaxation, and mediates XLF/LIG4 recruitment to DSBs for enhanced NHEJ efficiency.	[53, 54]
VPS35	Binding and isolating Ku70/80 away from the damage site, inhibiting DNA-PKcs activation, and downstream factor (XLF/LIG4) recruitment.	[55]

Table 1.
Accessory factors in the NHEJ pathway.

cell cycle. The critical determinant for pathway selection is whether DNA ends undergo resection, an essential prerequisite for HR [56]. Upon resection, the MRE11-Rad50-Nbs1 and CtBP-interacting protein (MRN-CtIP complex) first relieve the inhibitory effect of cell cycle and apoptosis regulator 2 (CCAR2) [22]. Relying on the endonuclease activity of MRE11, a gap of up to 300 bp from the 5' end is created. Subsequently, MRE11 or the exonuclease EXD2, utilizing 3'→5' exonuclease activity, extends the gap to the DSB end, accomplishing short-range resection [23, 57]. The resulting 3' single-stranded DNA (ssDNA) may then undergo extensive long-range resection mediated by EXO1 or the DNA2-BLM complex [58, 59]. DNA end resection directly influences DSB repair pathway choice, with several proteins modulating this process – most notably, p53-binding protein 1 (53BP1) and breast cancer type 1 susceptibility protein (BRCA1) [60]. As early as 2013, studies demonstrated that 53BP1-RIF1 and BRCA1-CtIP form a cell-cycle–dependent regulatory loop governing DSB repair choice: 53BP1-RIF1 inhibits end resection, thus promoting NHEJ, whereas BRCA1-CtIP facilitates end resection, thereby promoting HR (**Figure 3**) [61].

3.2.1. 53BP1 inhibits end resection and promotes NHEJ

Although 53BP1 lacks catalytic activity, its numerous DDR protein-interaction domains enable it to act as a pivotal recruitment platform during DSB repair [62, 63]. A landmark study in 2010 first proposed that 53BP1 could promote NHEJ by inhibiting CtIP-dependent end resection [64]. This hypothesis rapidly fueled research efforts to identify 53BP1 downstream effectors. Notably, research led by Jimeno and colleagues demonstrated that 53BP1 is recruited to DSB-proximal chromatin via histone methylation and ubiquitination, and that, when activated by ATM-mediated phosphorylation, it recruits RIF1 and its downstream Shieldin complex [65]. Within the Shieldin complex, Shieldin complex subunit 2 (SHLD2),

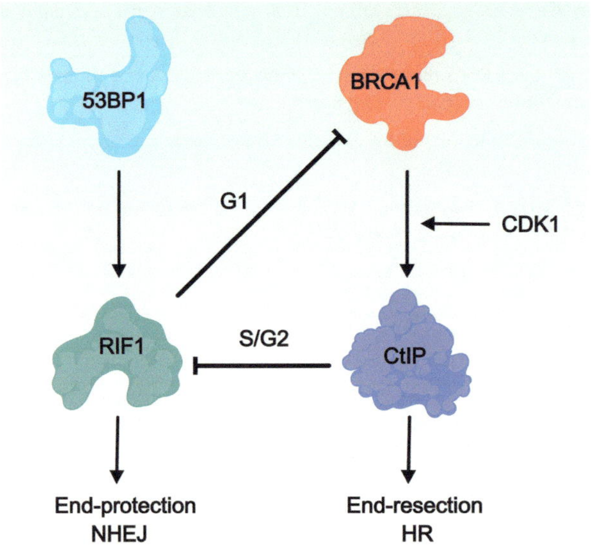

Figure 3.
53BP1-RIF1 and BRCA1-CTIP form a cell-cycle-dependent regulatory loop.

which harbors three OB-fold domains (oligonucleotide-binding fold), directly binds ssDNA to impede further EXO1-mediated resection [66]. Additionally, the Shieldin complex recruits the CTC1-STN1-TEN1 (CST) complex and the DNA polymerase α-primase complex (Polα-primase), with the latter able to reverse resection by fill-in synthesis, thus further facilitating NHEJ repair (**Figure 4**a) [67].

3.2.2. BRCA1 promotes end resection and enhances HR

BRCA1, an early-discovered tumor suppressor gene, exerts its formidable tumor-suppressive functions primarily by regulating DNA repair, promoting end resection, and enhancing HR. These functions are primarily mediated by its heterodimerization with BARD1 [68]. In addition to directly binding CtIP and facilitating MRN complex-mediated resection, BRCA1's key role in promoting end resection lies in abolishing the inhibitory effect of 53BP1. During the S/G2 phase, BRCA1 presence significantly reduces the size of 53BP1 foci, spatially segregating 53BP1 from DSB core regions – a process dependent on the chromatin remodeler SMARCAD1 [60, 69]. The BRCA1-BARD1 heterodimer catalyzes ubiquitination of histone H2A at lysine residues 125, 127, and 129, which enables SMARCAD1, via its CUE ubiquitin-binding domain, to be recruited to DSBs [68]. In yeast, the SMARCAD1 homologue Fun30 modulates 53BP1 localization at DSB cores via nucleosome sliding or H2A–H2B eviction, allowing access for resection machinery (**Figure 4**b) [70]. Beyond promoting end resection, BRCA1-BARD1 also facilitates RAD51 replacement and loading, thereby promoting HR-mediated repair [71].

4. NHEJ abnormalities and the occurrence of diseases

4.1. NHEJ and tumor

On the one hand, NHEJ's rapid repair capability stems from its ability to directly ligate broken DNA ends without relying on homologous templates. However, this error-prone process can lead to genomic alterations, such as gene deletions or chromosomal translocations, when repairing complex DNA lesions. Moreover, in cases of multiple DSBs, NHEJ may erroneously join ends from different chromosomes, further promoting chromosomal structural variations. These genetic aberrations provide a critical molecular foundation for tumor initiation and progression.

On the other hand, tumor cells often hyperactivate the NHEJ pathway to counteract DNA damage and sustain proliferation. Key NHEJ components (e.g., Ku70/80, DNA-PKcs, and LIG4) are frequently overexpressed in many cancers, which is related to tumor progression and poor prognosis [72, 73]. For example, DNA-PK inhibition enhances radiosensitivity and chemotherapeutic efficacy by inducing tumor cell senescence or apoptosis [74, 75]. Dysregulation of Artemis, LIG4, or XLF alters radiation responses. In HeLa cells, an Artemis mutation (D37N-413aa) stabilizes its interaction with DNA-PK, impairing endonuclease activity and NHEJ efficiency to promote apoptosis [76]. The acetyltransferase ESCO2 is abnormally highly expressed in samples from colorectal cancer patients. ESCO2 deletion increases the sensitivity of colorectal cancer cells to bleomycin, oxaliplatin, and other chemotherapy drugs [50]. Collectively, NHEJ components represent

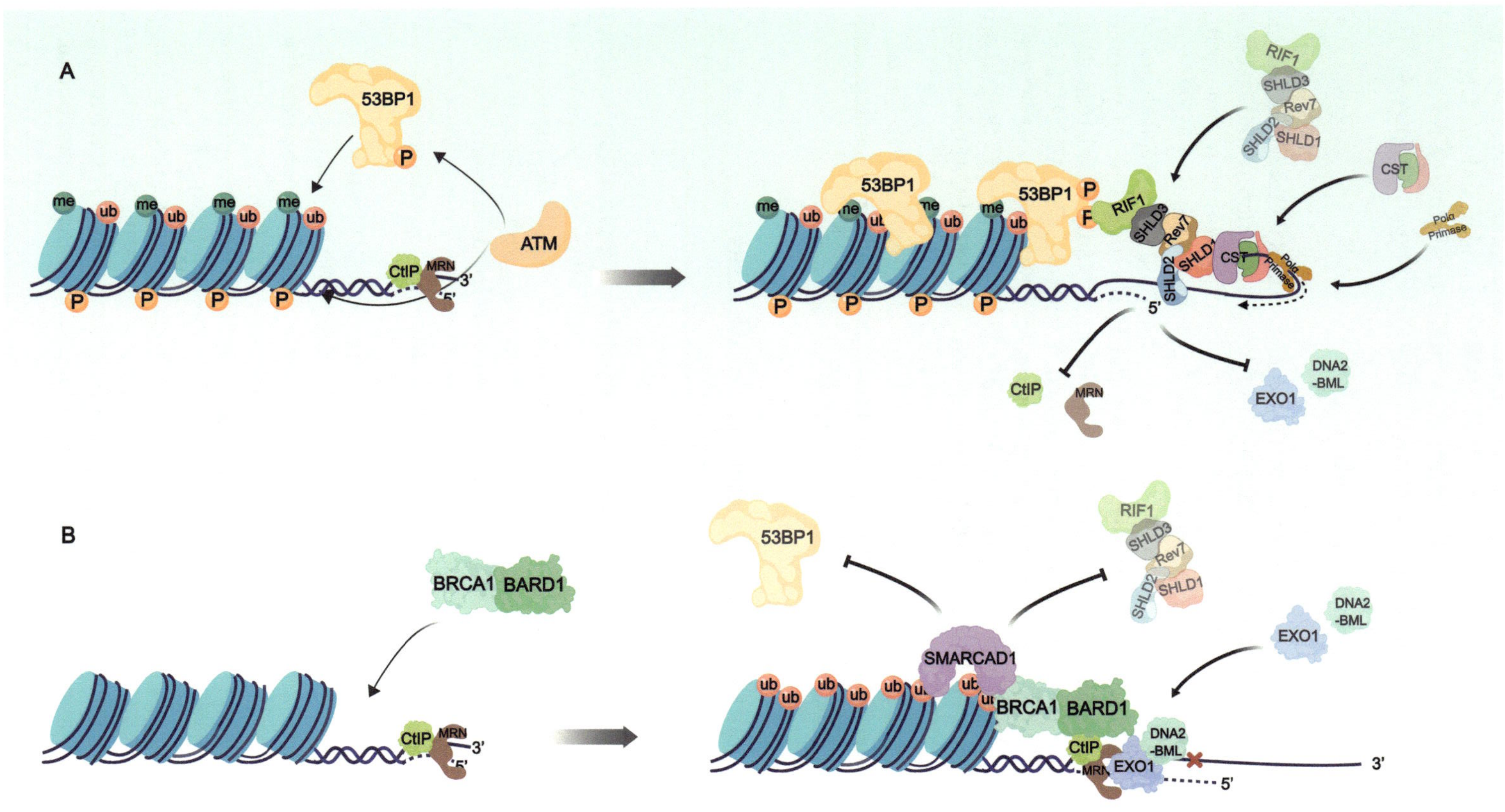

Figure 4.
Antagonistic regulation of DNA end resection by 53BP1 and BRCA1. (a)53BP1-mediated suppression of resection. 53BP1 is recruited to DSBs via histone methylation and ubiquitination modification. Following ATM-dependent phosphorylation, 53BP1 recruits RIF1 and its downstream Shieldin complex to protect the DSB end from excision by EXO1 and DNA2-BLM. Shieldin recruits CST and Polα/primase, promoting the fill-in reaction to counteract the DSB end resection. (b) BRCA1-mediated promotion of resection. BRCA1 binds to the CtIP-MRN complex and relies on the chromatin remodeling enzyme SMARCAD1 to isolate 53BP1 from the core region of DSBs, spatially allowing the entry of terminal resection proteins, thereby facilitating the resection process.

promising therapeutic targets, though their cell-type-specific roles warrant further investigation to optimize clinical strategies.

4.2. NHEJ and immune deficiency

V(D)J recombination is a physiological process that utilizes NHEJ to generate antigen receptor diversity through programmed DSBs during lymphocyte development [77]. This process occurs exclusively in developing B and T cells and consists of two distinct phases: DNA cleavage and repair. During the cleavage phase, the lymphocyte-specific RAG1/2 complex recognizes and cleaves flanking recombination signal sequences, creating coding ends with hairpin structures. In the repair phase, the Artemis-DNA-PKcs complex opens these hairpins, allowing processing and ligation of coding ends by the NHEJ machinery to complete chromosomal rejoining [78]. Defects in NHEJ components critically impair V(D)J recombination and lymphocyte development. DNA-PKcs mutations prevent effective coding end joining, disrupting T-cell receptor (TCR) and B-cell receptor (BCR) gene rearrangement and causing severe combined immunodeficiency with blocked lymphocyte maturation [79]. Artemis is a key endonuclease in the NHEJ pathway and is involved in the cleavage of DNA hairpin structures during V(D)J recombination. When the Artemis gene is mutated, the V(D)J recombination of TCR and BCR is blocked, which leads to the arrest of T/B lymphocyte development [80]. LIG4 is responsible for catalyzing the final connection step of the DNA ends in V(D)J recombination. LIG4 mutations impair V(D)J recombination by causing either loss or misconnection of recombination products, which compromises antigen receptor diversity. This molecular defect leads to T cell function defects and B cell antibody class switching disorders, resulting in combined immunodeficiency manifested as repeated bacterial infections and hypogammaglobulinemia [81]. These clinical manifestations underscore the essential role of NHEJ in establishing adaptive immune competence through V(D)J recombination.

4.3. NHEJ abnormality and neurodegenerative diseases

Neurons' high metabolic activity, primarily dependent on oxidative phosphorylation, increases reactive oxygen species generation and subsequent DNA damage, rendering them exceptionally vulnerable. Therefore, efficient and timely DNA damage repair is critical for genomic integrity and neuronal survival. The nervous system depends on efficient NHEJ-mediated DNA repair to maintain genomic stability, particularly during neuronal differentiation [82]. This dependence on NHEJ is clinically evident in ALS, a rapidly progressive neurodegenerative disorder characterized by the degeneration of upper and lower motor neurons, leading to widespread muscle atrophy [83, 84]. Genetic studies have identified over 40 ALS-associated genes, with TARDBP (encoding TDP-43) and FUS emerging as particularly significant for their dual roles in DNA repair and disease pathogenesis [85, 86]. In healthy neurons, TDP-43 facilitates NHEJ-mediated repair of DSBs by promoting the nuclear translocation and recruitment of the XRCC4-DNA LIG4 complex to damage sites, forming complexes with key repair factors such as Ku, DNA-PKcs, and 53BP1. However, the ALS-associated TDP-43 mutation (Q331K) disrupts this process, preventing XRCC4-DNA Ligase 4 nuclear translocation and leading to persistent DSBs, which trigger DNA-damage-mediated neuronal apoptosis [87]. Similarly, the

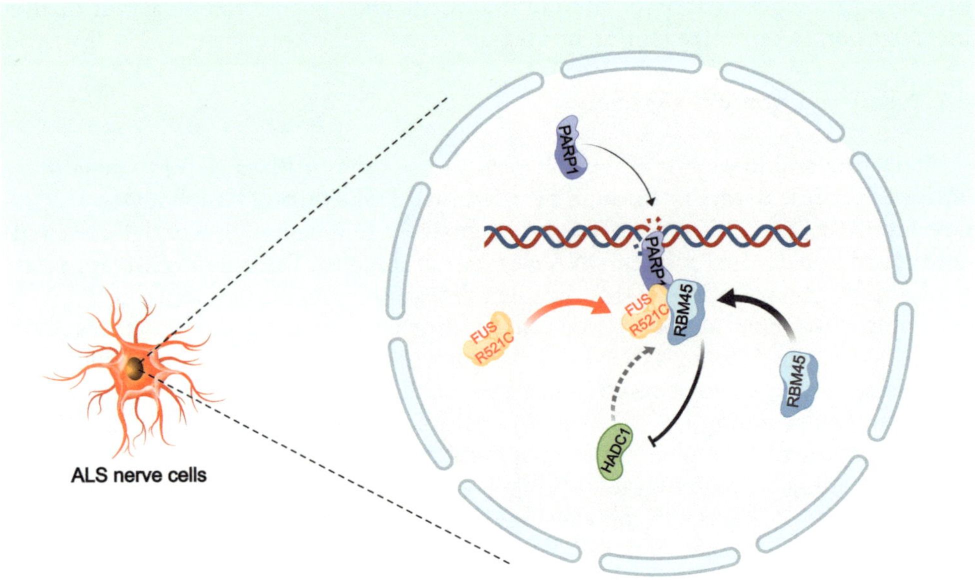

Figure 5.
Model of RBM45 in DNA damage response under ALS mutation conditions. FUS-R521C, a mutant found in fALS patients, enhances its interaction with RBM45, which decreases RBM45's association with HDAC1 and diminishes the recruitment of HDAC1.

ALS-linked FUS mutation (R521C) enhances its interaction with RBM45, which significantly weakens the mutant FUS's interaction with histone deacetylase 1 (HDAC1). This weakened interaction impairs HDAC1 recruitment to DNA damage sites and reduces NHEJ repair efficiency (**Figure 5**), thereby decreasing genomic stability and contributing to neuronal functional degeneration and death [88–90].

5. NHEJ-targeted therapy and application

Recent advances in targeting the NHEJ pathway have yielded significant breakthroughs across multiple therapeutic approaches, particularly in synthetic lethality and precision medicine interventions. In particular, for tumors with TP53 mutations, the DNA-PKcs inhibitor M3841 is used in combination with the POL θ inhibitor novobiocin to simultaneously block the NHEJ and alt-NHEJ pathways, resulting in the accumulation of DNA DSB end-resection toxicity, thereby inducing synthetic lethality [91]. In non-small-cell lung cancer, a new class of DNA-PK inhibitors can block the interaction between Ku70/80 heterodimers and DNA, affecting the repair efficiency of NHEJ, thereby enhancing the sensitivity of cells to bleomycin and ionizing radiation [92]. SIRT6, as an important aging regulatory protein, can participate in HR and NHEJ repair by targeting proteins such as PARP1 and DNA-PKcs [93], while a new SIRT6 allosteric activator, MDL-800, can activate the activity of mouse SIRT6 deacetylase to promote the repair efficiency of NHEJ and maintain the stability of the genome, providing iPSC-based cell therapy for the treatment of aging-related diseases a new direction [94]. These approaches demonstrate the expanding therapeutic potential of NHEJ modulation, and further elucidation of NHEJ mechanisms promises to unlock additional targeted treatment opportunities.

6. Future perspectives: Challenges and opportunities in NHEJ research

The dual nature of NHEJ, as both an essential DNA repair pathway and a potential source of genomic instability, presents unique research challenges and therapeutic opportunities. Currently, research into the mechanisms of NHEJ repair stands at the intersection of emerging opportunities and significant challenges. On the technical level, ongoing advancements in cutting-edge technologies, such as cryo-electron microscopy and single-molecule imaging, enable researchers to dissect the dynamic assembly of NHEJ repair protein complexes at atomic resolution, thereby profoundly unraveling the molecular mechanisms underlying its repair processes. Meanwhile, the deep integration of artificial intelligence and machine learning algorithms holds promise for accurately predicting the preference and error-proneness of NHEJ repair under different cellular environments, breaking through the current bottleneck of understanding its complex regulatory network. In terms of applications, as a critical player in gene editing technology, the precise regulation of NHEJ repair may become a key breakthrough in gene therapy, offering new avenues for combating major diseases such as genetic disorders and cancer. Additionally, the development of drugs specifically targeting the NHEJ pathway holds potential for advancing novel anti-tumor targeted therapies, which may enhance tumor therapeutic efficacy while minimizing adverse side effects. Furthermore, exploring the mechanistic role of NHEJ repair in aging and neurodegenerative diseases may uncover innovative strategies for alleviating aging processes and managing neurological disorders. The convergence of these multidisciplinary approaches is poised to redefine the boundaries of DNA repair research and catalyze breakthroughs in treating genetic disorders, cancer, and age-related diseases.

7. Conclusions

NHEJ represents a sophisticated, yet error-prone DNA repair system that plays dual roles in preserving genomic stability while simultaneously contributing to disease pathogenesis. This essential pathway, when dysregulated, becomes a driving force behind oncogenic transformation through genomic instability, immunodeficiency disorders due to defective V(D)J recombination, and progressive neurodegeneration. The mechanistic understanding of NHEJ has revealed promising therapeutic targets, with DNA-PKcs inhibitors and synthetic lethality approaches demonstrating significant potential in oncology, while strategies to enhance repair fidelity or address neuron-specific deficiencies may transform treatment paradigms for aging-related and neurodegenerative conditions. Emerging technologies in structural biology and computational modeling are poised to unravel the nuanced regulation of NHEJ across cellular contexts, paving the way for targeted therapeutic innovations that address its multifaceted roles in human health and disease.

Acknowledgments

This work was funded by the Education Department Foundation of Heilongjiang Province (2024-KYYWF-0339), the Construction Project of Dominant Characteristic Disciplines of Qiqihar Medical University (QYZDXK-003), and the Qiqihar Medical Institute Foundation (QMSI2024Z-01).

Conflict of interest

The authors declare no conflict of interest.

Author details

Yan-Ji Lu[1,2], Xian Hong[1] and Tao Wang[1,2]*

1 Laboratory of Protein Structure and Function, Institute of Medicine and Pharmacy, Qiqihar Medical University, Qiqihar, China

2 School of Medical Technology, Qiqihar Medical University, Qiqihar, China

*Address all correspondence to: wangtao@qmu.edu.cn

References

[1] O'Driscoll M. Diseases associated with defective responses to DNA damage. Cold Spring Harbor Perspectives in Biology. 2012;**4**(12):a012773. DOI: 10.1101/cshperspect.a012773

[2] Burlet D, Khan MM, Hacot S, et al. The inflammasome sensor NLRP3 interacts with REV7 to maintain genome integrity through homologous recombination. Nucleic Acids Research. 2025;**53**(12):gkaf554. DOI: 10.1093/nar/gkaf554.

[3] Zhang -W-W, Wright DG, Harrison L, Matlashewski G, Blader IJ. 2022. Reconstitution of Mycobacterium marinum nonhomologous DNA end joining pathway in Leishmania. mSphere. 7(3):e00156–22. DOI: 10.1128/msphere.00156-22

[4] Loparo JJ. Holding it together: DNA end synapsis during non-homologous end joining. DNA Repair (Amst). 2023;**130**:103553. DOI: 10.1016/j.dnarep.2023.103553.

[5] Mikhova M, Goff NJ, Janovič T, Heyza JR, Meek K, Schmidt JC. 2024. Single-molecule imaging reveals the kinetics of non-homologous end-joining in living cells. Nature Communications. **15**(1):10159. DOI: 10.1038/s41467-024-54545-y.

[6] Mao Z, Bozzella M, Seluanov A, Gorbunova V. 2008. Comparison of nonhomologous end joining and homologous recombination in human cells. DNA Repair (Amst) 7(10):1765–1771. DOI: 10.1016/j.dnarep.2008.06.018.

[7] Beucher A, Birraux J, Tchouandong L, et al. ATM and Artemis promote homologous recombination of radiation-induced DNA double-strand breaks in G2. The EMBO Journal. 2009;**28**(21):3413–3427. DOI: 10.1038/emboj.2009.276.

[8] Yang K, Guo R, Xu D. 2016. Non-homologous end joining: Advances and frontiers. Acta Biochimica Et Biophysica Sinica. **48**(7):632–640. DOI: 10.1093/abbs/gmw046

[9] Yang G, Liu C, Chen SH, et al. Super-resolution imaging identifies PARP1 and the Ku complex acting as DNA double-strand break sensors. Nucleic Acids Research. 2018;**46**(7):3446–3457. DOI: 10.1093/nar/gky088.

[10] Downs JA, Jackson SP. 2004. A means to a DNA end: The many roles of Ku. Nature Reviews Molecular Cell Biology. **5**(5):367–378. DOI: 10.1038/nrm1367

[11] Mimori T, Hardin JA. Mechanism of interaction between Ku protein and DNA. Journal of Biological Chemistry. 1986;**261**(22):10375–10379

[12] Yaneva M. 1997. Interaction of DNA-dependent protein kinase with DNA and with Ku: Biochemical and atomic-force microscopy studies. The EMBO Journal. **16**(16):5098–5112. DOI: 10.1093/emboj/16.16.5098

[13] Nick Mcelhinny SA, Snowden CM, McCarville J, Ramsden DA. 2000. Ku recruits the XRCC4-ligase IV complex to DNA ends. Molecular & Cellular Biology. **20**(9):2996–3003. DOI: 10.1128/mcb.20.9.2996-3003.2000

[14] Chen L, Trujillo K, Sung P, Tomkinson AE. 2000. Interactions of

the DNA ligase IV-XRCC4 complex with DNA ends and the DNA-dependent protein kinase. Journal of Biological Chemistry. **275**(34):26196–26205. DOI: 10.1074/jbc.M000491200

[15] Lieber MR. The mechanism of double-strand DNA break repair by the nonhomologous DNA end-joining pathway. Annual Review of Biochemistry. 2010;**79**:181–211. DOI: 10.1146/annurev.biochem.052308.093131.

[16] Abraham RT. 2004. PI 3-kinase related kinases: 'Big' players in stress-induced signaling pathways. DNA Repair (Amst.). **3**(8-9):883–887. DOI: 10.1016/j.dnarep.2004.04.002

[17] Ding Q, Reddy YV, Wang W, et al. Autophosphorylation of the catalytic subunit of the DNA-dependent protein kinase is required for efficient end processing during DNA double-strand break repair. Molecular & Cellular Biology. 2003;**23**(16):5836–5848. DOI: 10.1128/mcb.23.16.5836-5848.2003.

[18] Gell D, Jackson SP. 1999. Mapping of protein-protein interactions within the DNA-dependent protein kinase complex. Nucleic Acids Research. **27**(17):3494–3502. DOI: 10.1093/nar/27.17.3494

[19] West RB, Yaneva M, Lieber MR. 1998. Productive and nonproductive complexes of Ku and DNA-dependent protein kinase at DNA termini. Molecular & Cellular Biology. **18**(10):5908–5920. DOI: 10.1128/mcb.18.10.5908.

[20] Yoo S. 1999. Geometry of a complex formed by double strand break repair proteins at a single DNA end: Recruitment of DNA-PKcs induces inward translocation of Ku protein. Nucleic Acids Research. **27**(24):4679–4686. DOI: 10.1093/nar/27.24.4679.

[21] Calsou P, Frit P, Humbert O, Muller C, Chen DJ, Salles B. 1999. The DNA-dependent protein kinase catalytic activity regulates DNA end processing by means of Ku entry into DNA. Journal of Biological Chemistry. **274**(12):7848–7856. DOI: 10.1074/jbc.274.12.7848.

[22] López-Saavedra A, Gómez-Cabello D, Domínguez-Sánchez MS, et al. A genome-wide screening uncovers the role of CCAR2 as an antagonist of DNA end resection. Nature Communications. 2016;**7**:12364.

[23] Garcia V, Phelps SE, Gray S, Neale MJ. 2011. Bidirectional resection of DNA double-strand breaks by Mre11 and Exo1. Nature. **479**(7372):241–244. DOI: 10.1038/nature10515

[24] Bebenek K, Pedersen LC, Kunkel TA. 2014. Structure–function studies of DNA polymerase λ. Biochemistry. **53**(17):2781–2792. DOI: 10.1021/bi4017236.

[25] Moon AF, Pryor JM, Ramsden DA, Kunkel TA, Bebenek K, Pedersen LC. 2014. Sustained active site rigidity during synthesis by human DNA polymerase μ. Nature Structural and Molecular Biology. **21**(3):253–260. DOI: 10.1038/nsmb.2766.

[26] Ma Y, Lu H, Tippin B, et al. A biochemically defined system for mammalian nonhomologous DNA end joining. Molecular Cell. 2004;**16**(5):701–713. DOI: 10.1016/j.molcel.2004.11.017.

[27] Bertocci B, De Smet A, Weill JC, Reynaud CA. 2006. Nonoverlapping functions of DNA polymerases mu, lambda, and terminal deoxynucleotidyl

transferase during immunoglobulin V(D)J recombination in vivo. Immunity. **25**(1):31–41. DOI: 10.1016/j.immuni.2006.04.013

[28] Sallmyr A, Rashid I, Bhandari SK, Naila T, Tomkinson AE. Human DNA ligases in replication and repair. DNA Repair (Amst). 2020;**93**:102908. DOI: 10.1016/j.dnarep.2020.102908.

[29] Liu L, Li J, Cisneros-Aguirre M, et al. Dynamic assemblies and coordinated reactions of non-homologous end joining. Nature. 2025. DOI: 10.1038/s41586-025-09078-9.

[30] Stinson BM, Carney SM, Walter JC, Loparo JJ. 2024. Structural role for DNA ligase IV in promoting the fidelity of non-homologous end joining. Nature Communications. **15**(1):1250. DOI: 10.1038/s41467-024-45553-z

[31] Ghosh D, Raghavan SC. 2021. Nonhomologous end joining: New accessory factors fine tune the machinery. Trends in Genetics. **37**(6):582–599. DOI: 10.1016/j.tig.2021.03.001

[32] Ochi T, Blackford AN, Coates J, et al. DNA repair. PAXX, a paralog of XRCC4 and XLF, interacts with Ku to promote DNA double-strand break repair. Science. 2015;**347**(6218):185–188. DOI: 10.1126/science.1261971.

[33] Liu X, Shao Z, Jiang W, Lee BJ, Zha S. PAXX promotes KU accumulation at DNA breaks and is essential for end-joining in XLF-deficient mice. Nature Communications. 2017;**8**:13816. DOI: 10.1038/ncomms13816.

[34] Tadi SK, Tellier-Lebègue C, Nemoz C, et al. PAXX is an accessory c-NHEJ factor that associates with Ku70 and has overlapping functions with XLF. Cell Reports. 2016;**17**(2):541–555. DOI: 10.1016/j.celrep.2016.09.026.

[35] Seif-El-Dahan M, Kefala-Stavridi A, Frit P, et al. PAXX binding to the NHEJ machinery explains functional redundancy with XLF. Science Advances. 2023;**9**(22):eadg2834. DOI: 10.1126/sciadv.adg2834.

[36] Hung PJ, Johnson B, Chen BR, et al. MRI is a DNA damage response adaptor during classical non-homologous end joining. Molecular Cell. 2018;**71**(2):332–42.e8. DOI: 10.1016/j.molcel.2018.06.018.

[37] Arnoult N, Correia A, Ma J, et al. Regulation of DNA repair pathway choice in S and G2 phases by the NHEJ inhibitor CYREN. Nature. 2017;**549**(7673):548–552. DOI: 10.1038/nature24023.

[38] Guerrero EN, Mitra J, Wang H, et al. Amyotrophic lateral sclerosis-associated TDP-43 mutation Q331K prevents nuclear translocation of XRCC4-DNA ligase 4 complex and is linked to genome damage-mediated neuronal apoptosis. Human Molecular Genetics. 2019;**28**(18):3161–3162. DOI: 10.1093/hmg/ddz141.

[39] Mitra J, Guerrero EN, Hegde PM, et al. Motor neuron disease-associated loss of nuclear TDP-43 is linked to DNA double-strand break repair defects. P Natl Acad Sci USA. 2019;**116**(10):4696–4705. DOI: 10.1073/pnas.1818415116.

[40] Konopka A, Whelan DR, Jamali MS, et al. Impaired NHEJ repair in amyotrophic lateral sclerosis is associated with TDP-43 mutations. Molecular Neurodegeneration. 2020;**15**(1):51. DOI: 10.1186/s13024-020-00386-4.

[41] Pryor JM, Conlin MP, Carvajal-Garcia J, et al. Ribonucleotide incorporation enables repair of chromosome breaks by nonhomologous end joining. Science. 2018;**361**(6407):1126–1129. DOI: 10.1126/science.aat2477.

[42] Li W, Bai X, Li J, et al. The nucleoskeleton protein IFFO1 immobilizes broken DNA and suppresses chromosome translocation during tumorigenesis. Nature Cell Biology. 2019;**21**(10):1273–1285. DOI: 10.1038/s41556-019-0388-0.

[43] Olivieri M, Cho T, Álvarez-Quilón A, et al. A genetic map of the response to DNA damage in human cells. Cell. 2020;**182**(2):481–96.e21. DOI: 10.1016/j.cell.2020.05.040.

[44] Liu X, Liu T, Shang Y, et al. ERCC6L2 promotes DNA orientation-specific recombination in mammalian cells. Cell Research. 2020;**30**(9):732–744. DOI: 10.1038/s41422-020-0328-3.

[45] Francica P, Mutlu M, Blomen VA, et al. Functional radiogenetic profiling implicates ERCC6L2 in non-homologous end joining. Cell Reports. 2020;**32**(8):108068. DOI: 10.1016/j.celrep.2020.108068.

[46] Jin H, Lee B, Luo Y, et al. FOXL2 directs DNA double-strand break repair pathways by differentially interacting with Ku. Nature Communications. 2020;**11**(1):2010. DOI: 10.1038/s41467-020-15748-1.

[47] Zhang H, Jiang L, Du X, et al. The cGAS-Ku80 complex regulates the balance between two end joining subpathways. Cell Death and Differentiation. 2024;**31**(6):792–803. DOI: 10.1038/s41418-024-01296-4.

[48] Bao K, Ma Y, Li Y, et al. A di-acetyl-decorated chromatin signature couples liquid condensation to suppress DNA end synapsis. Molecular Cell. 2024;**84**(7):1206–23.e15. DOI: 10.1016/j.molcel.2024.02.002.

[49] Singh JK, Smith R, Rother MB, et al. Zinc finger protein ZNF384 is an adaptor of Ku to DNA during classical non-homologous end-joining. Nature Communications. 2021;**12**(1):6560. DOI: 10.1038/s41467-021-26691-0.

[50] Fu J, Zhou S, Xu H, et al. ATM–ESCO2–SMC3 axis promotes 53BP1 recruitment in response to DNA damage and safeguards genome integrity by stabilizing cohesin complex. Nucleic Acids Research. 2023;**51**(14):7376–7391. DOI: 10.1093/nar/gkad533.

[51] Gomez TS, Billadeau DD. 2009. A FAM21-containing WASH complex regulates retromer-dependent sorting. Developmental Cell. **17**(5):699–711. DOI: 10.1016/j.devcel.2009.09.009

[52] Derivery E, Sousa C, Gautier JJ, Lombard B, Loew D, Gautreau A. 2009. The Arp2/3 activator WASH controls the fission of endosomes through a large multiprotein complex. Developmental Cell. **17**(5):712–723. DOI: 10.1016/j.devcel.2009.09.010

[53] Wang T, Du XH, Hong Y, et al. WASH interacts with Ku to regulate DNA double-stranded break repair. iScience. 2022;**25**(1):103676. DOI: 10.1016/j.isci.2021.103676.

[54] Wang T, Zheng AX, Li P, et al. FAM21 interacts with Ku to promote the localization of WASH to DNA double strand break sites. DNA Repair (Amst.). 2024;**133**:103603

[55] Zhang L, Nie Y, Tang T, et al. Nuclear VPS35 attenuates NHEJ repair by sequestering Ku protein. Molecular Medicine. 2025;**31**(1):222. DOI: 10.1186/s10020-025-01288-1.

[56] Biehs R, Steinlage M, Barton O, et al. DNA double-strand break resection occurs during non-homologous end joining in G1 but is distinct from resection during homologous recombination. Molecular Cell. 2017;**65**(4):671–84.e5. DOI: 10.1016/j.molcel.2016.12.016.

[57] Broderick R, Nieminuszczy J, Baddock HT, et al. EXD2 promotes homologous recombination by facilitating DNA end resection. Nature Cell Biology. 2016;**18**(3):271–280. DOI: 10.1038/ncb3303.

[58] Mimitou EP, Symington SLS. 2008. Sae2, Exo1 and Sgs1 collaborate in DNA double-strand break processing. Nature. **455**(7214):770–774. DOI: 10.1038/nature07312

[59] Nimonkar AV, Genschel J, Kinoshita E, et al. BLM–DNA2–RPA–MRN and EXO1–BLM–RPA–MRN constitute two DNA end resection machineries for human DNA break repair. Genes and Development. 2011;**25**(4):350–362. DOI: 10.1101/gad.2003811.

[60] Daley JM, Sung P. 2014. 53BP1, BRCA1, and the choice between recombination and end joining at DNA double-strand breaks. Molecular & Cellular Biology. **34**(8):1380–1388. DOI: 10.1128/mcb.01639-13

[61] Escribano-Díaz C, Orthwein A, Fradet-Turcotte A, et al. A cell cycle-dependent regulatory circuit composed of 53BP1-RIF1 and BRCA1-CtIP controls DNA repair pathway choice. Molecular Cell. 2013;**49**(5):872–883. DOI: 10.1016/j.molcel.2013.01.001.

[62] Dai Y, Zhang A, Shan S, Gong Z, Zhou Z. 2018. Structural basis for recognition of 53BP1 tandem Tudor domain by TIRR. Nature Communications. **9**(1):2123. DOI: 10.1038/s41467-018-04557-2.

[63] Chapman JR, Sossick AJ, Boulton SJ, Jackson SP. 2012. BRCA1-associated exclusion of 53BP1 from DNA damage sites underlies temporal control of DNA repair. Journal of Cell Science. **125**(15):3529–3534. DOI: 10.1242/jcs.105353

[64] Bunting SF, Callén E, Wong N, et al. 53BP1 inhibits homologous recombination in Brca1-deficient cells by blocking resection of DNA breaks. Cell. 2010;**141**(2):243–254. DOI: 10.1016/j.cell.2010.03.012.

[65] Jimeno S, Mejías-Navarro F, Prados-Carvajal R, Huertas P. Controlling the balance between chromosome break repair pathways. Advances in Protein Chemistry and Structural Biology 2019;**115**:95–134. DOI: 10.1016/bs.apcsb.2018.10.004.

[66] Setiaputra D, Durocher D. Shieldin–the protector of DNA ends. EMBO Reports. 2019;**20**(5). DOI: 10.15252/embr.201847560.

[67] Mirman Z, Lottersberger F, Takai H, et al. 53BP1–RIF1–shieldin counteracts DSB resection through CST- and Polα-dependent fill-in. Nature. 2018;**560**(7716):112–116. DOI: 10.1038/s41586-018-0324-7.

[68] Densham RM, Garvin AJ, Stone HR, et al. Human BRCA1–BARD1 ubiquitin ligase activity counteracts chromatin barriers to DNA resection. Nature Structural and Molecular

Biology. 2016;**23**(7):647–655. DOI: 10.1038/nsmb.3236.

[69] Densham RM, Morris JR. Moving mountains—The BRCA1 promotion of DNA resection. Frontiers in Molecular Biosciences. 2019;**6**:79. DOI: 10.3389/fmolb.2019.00079.

[70] Awad S, Ryan D, Prochasson P, Owen-Hughes T, Hassan AH. 2010. The Snf2 homolog Fun30 acts as a homodimeric ATP-dependent chromatin-remodeling enzyme. Journal of Biological Chemistry. **285**(13):9477–9484. DOI: 10.1074/jbc.M109.082149

[71] Tarsounas M, Sung P. 2020. The antitumorigenic roles of BRCA1–BARD1 in DNA repair and replication. Nature Reviews Molecular Cell Biology. **21**(5):284–299. DOI: 10.1038/s41580-020-0218-z

[72] Katheeja MN, Das SP, Das R, Laha S. BRCA1 interactors, RAD50 and BRIP1, as prognostic markers for triple-negative breast cancer severity. Frontiers in Genetics. 2023;**14**:1035052. DOI: 10.3389/fgene.2023.1035052.

[73] Lin Z, Huang Z, Shi Y, et al. A novel NHEJ gene signature based model for risk stratification and prognosis prediction in hepatocellular carcinoma. Cancer Cell International. 2023;**23**(1):59. DOI: 10.1186/s12935-023-02907-9.

[74] Kovina AP, Luzhin AV, Tatarskiy VV, et al. Disruption of RNA splicing increases vulnerability of cells to DNA-PK inhibitors. International Journal of Molecular Sciences. 2024;**25**(21):11810. DOI: 10.3390/ijms252111810.

[75] Wang W, McMillan MT, Zhao X, et al. DNA-PK inhibition and radiation promote antitumoral immunity through RNA polymerase III in pancreatic cancer. Molecular Cancer Research. 2022;**20**(7):1137–1150. DOI: 10.1158/1541-7786.Mcr-21-0725.

[76] Liu H, Sun X, Zhang S, Ge W, Zhu Y, Zhang J, Zheng S, et al. The dominant negative mutant Artemis enhances tumor cell radiosensitivity. Radiotherapy & Oncology. 2011;**101**(1):66–72. DOI: 10.1016/j.radonc.2011.05.034.

[77] Strande NT, Waters CA, Ramsden DA. 2012. Resolution of complex ends by nonhomologous end joining - better to be lucky than good? Genome Integrity. **3**(1):10. DOI: 10.1186/2041-9414-3-10.

[78] Ma Y, Pannicke U, Schwarz K, Lieber MR. 2002. Hairpin opening and overhang processing by an Artemis/DNA-dependent protein kinase complex in nonhomologous end joining and V(D)J recombination. Cell. **108**(6):781–794. DOI: 10.1016/s0092-8674(02)00671-2

[79] Adelon J, Abolhassani H, Esenboga S, et al. Human DNA-dependent protein kinase catalytic subunit deficiency: A comprehensive review and update. Journal of Allergy and Clinical Immunology. 2024;**154**(5):1300–1312. DOI: 10.1016/j.jaci.2024.06.018.

[80] Bin Y, Wei S, Chen R, et al. Dclre1c-mutation-induced immunocompromised mice are a novel model for human xenograft research. Biomolecules. 2024;**14**(2):180. DOI: 10.3390/biom14020180.

[81] Jauch AJ, Bignucolo O, Seki S, et al. Autoimmunity and immunodeficiency associated with monoallelic LIG4 mutations via haploinsufficiency. Journal of Allergy and Clinical

Immunology. 2023;**152**(2):500–516. DOI: 10.1016/j.jaci.2023.03.022.

[82] Orii KE, Lee Y, Kondo N, McKinnon PJ. 2006. Selective utilization of nonhomologous end-joining and homologous recombination DNA repair pathways during nervous system development. Proceedings of the National Academy of Sciences of the United States of America. **103** (26):10017–10022. DOI: 10.1073/pnas.0602436103.

[83] Yu B, Pamphlett R. Environmental insults: Critical triggers for amyotrophic lateral sclerosis. Translational Neurodegeneration. 2017;**6**:15. DOI: 10.1186/s40035-017-0087-3.

[84] Ragagnin AMG, Shadfar S, Vidal M, Jamali MS, Atkin JD. Motor neuron susceptibility in ALS/FTD.. Frontiers in Neuroscience. 2019;**13**:532. DOI: 10.3389/fnins.2019.00532.

[85] Guerrero EN, Wang H, Mitra J, et al. TDP-43/FUS in motor neuron disease: Complexity and challenges.. Progress in Neurobiology. 2016;**145-146**:78–97

[86] Peters OM, Ghasemi M, Brown Jr RH. 2015. Emerging mechanisms of molecular pathology in ALS. Journal of Clinical Investigation. **125**(6):2548. DOI: 10.1172/jci82693.

[87] Wang H, Kodavati M, Britz GW, Hegde ML. DNA damage and repair deficiency in ALS/FTD-associated neurodegeneration: From molecular mechanisms to therapeutic implication. Frontiers in Molecular Neuroscience. 2021;**14**:784361. DOI: 10.3389/fnmol.2021.784361.

[88] Gupta S, You P, SenGupta T, Nilsen H, Sharma K. Crosstalk between different DNA repair pathways contributes to neurodegenerative diseases.. Biology (Basel). 2021;**10**(2). DOI: 10.3390/biology10020163.

[89] Ding Q, Chaplin J, Morris MJ, et al. TDP-43 mutation affects stress granule dynamics in differentiated NSC-34 motoneuron-like cells. Frontiers in Cell and Developmental Biology. 2021;**9**:611601.

[90] Gong J, Huang M, Wang F, et al. RBM45 competes with HDAC1 for binding to FUS in response to DNA damage. Nucleic Acids Research. 2017;**45**(22):12862–12876. DOI: 10.1093/nar/gkx1102.

[91] Patterson-Fortin J, Bose A, Tsai WC, et al. Targeting DNA repair with combined inhibition of NHEJ and MMEJ induces synthetic lethality in TP53-mutant cancers. Cancer Research. 2022;**82**(20):3815–3829. DOI: 10.1158/0008-5472.Can-22-1124.

[92] Mendoza-Munoz PL, Gavande NS, VanderVere-Carozza PS, et al. Ku–DNA binding inhibitors modulate the DNA damage response in response to DNA double-strand breaks. NAR Cancer. 2023;**5**(1):zcad003. DOI: 10.1093/narcan/zcad003.

[93] Affandi T, Haas A, Ohm AM, Wright GM, Black JC, Reyland ME. 2024. PKC δ regulates chromatin remodeling and DNA repair through SIRT6. Molecular Cancer Research. **22**(2):181–196. DOI: 10.1158/1541-7786.Mcr-23-0493.

[94] Chen Y, Chen J, Sun X, Yu J, Qian Z, Wu L, Xu X, Wan X, Jiang Y, Zhang J, Gao S, Mao Z, et al. The SIRT6 activator MDL-800 improves genomic stability and pluripotency of old murine-derived iPS cells. Aging Cell. 2020;**19**(8):e13185. DOI: 10.1111/acel.13185.

Section 3

Epigenetics and Inheritance: Mechanisms and Metabolic Regulation

Chapter 5

Epigenetic Inheritance: Maintaining and Fine-Tuning Origin-Specific Features Across Generations

Jiabin Duan, Md Abdur Razzaque and Shaohua Qi

Abstract

Epigenetic inheritance refers to the transmission of gene regulatory information across generations without alterations in the underlying DNA sequence. Among its most striking manifestations is the faithful maintenance of parent-of-origin-specific expression – and, in some cases, the nuanced refinement – of origin-specific features from parent to offspring, wherein certain gene expressions are maintained preferentially from either the maternal or paternal allele, a phenomenon primarily governed by genomic imprinting. Central to this phenomenon are epigenetic modifications, including DNA methylation, histone modifications, non-coding RNAs, and chromatin remodelers, which together preserve developmental programs, imprinting patterns, and stress responses across cellular and organismal generations. Additionally, environmental cues experienced by parents can reshape epigenomic landscapes and influence offspring phenotypes through transgenerational epigenetic inheritance. These heritable marks are not merely static carriers of information but are dynamically regulated to alter gene expression in response to developmental context and external demands. Understanding how epigenetic memory is selectively maintained, erased, or modified holds profound implications for human health, evolution, and the treatment of diseases caused by epigenomic dysregulation. In this review, we interpret the mechanisms underlying the maintenance of these origin-specific features in relation to the regulation and preservation of epigenetic marks during germline development and early embryogenesis. We discuss how some of these marks escape epigenetic reprogramming, enabling the faithful inheritance of regulatory states. Additionally, we highlight the role of epigenetic programming, imprinting disorders, and the effect of environmental factors in shaping transgenerational phenotypes. A deeper understanding of epigenetic inheritance provides valuable insights and holds promise for future therapeutic strategies aimed at correcting heritable epigenetic dysregulation.

Keywords: epigenetic inheritance, genetic features, epigenetic reprogramming, inherited diseases, therapeutic strategies

1. Introduction

Traditionally, the inheritance of biological traits has been understood through the transmission of DNA sequence information from one generation to the next. However, heritable changes in gene expression can also be finely controlled independently of DNA sequence alterations. The epigenetic marks include reversible modifications of DNA and histones [1, 2]. This concept challenges the classical Mendelian view of heredity by demonstrating that environmental cues, developmental history, and parental conditions can leave molecular "imprints" that persist across generations. Unlike stereotypical genetic inheritance, epigenetic inheritance involves the transmission of gene regulatory information through mechanisms such as DNA methylation, histone modifications, non-coding RNAs, and chromatin remodeling, all of which influence chromatin accessibility and gene expression activity. These modification enzymes are dynamic, yet under certain circumstances, they can be stably maintained during mitotic and meiotic divisions, allowing regulatory states to be passed on to offspring.

Epigenetic mechanisms represent an intriguing and intricate conception of various heritable changes in gene expression features that are independent of changes in genetic material and reprogram genomic information. Evidence from both animal models and human populations has suggested that epigenetic modifications participate in a broad spectrum of processes. In the case of inheritance, the DNA code that passes from ancestors to offspring is reprogrammed by parents' experiences in the form of epigenetic tags. Genetic information from eggs and sperm is not erased to a blank state; a small minority of genes retains these features and pass them to the next generation [3]. The related genetic traits and plastic epigenetic features are inherited, including skill ability, emotional preference, lifestyle, and tumor initiation and resistance.

Epigenetic modifications are carried out through the "fine-tuning" processes to maintain the homeostasis of a "healthy" condition. Once the "fine-tuning" processes are disrupted, the levels of DNA replication, gene transcription, chromatin structure, and chromosome stability will also alter accordingly, ultimately resulting in epigenetic changes in germ cells that are delivered to the next generation, influencing disease susceptibility and phenotype. Preclinical and clinical findings demonstrate that dysfunctions in epigenetic regulation during these stages can lead to developmental aberrations and long-term health consequences, especially familial diseases. The underlying mechanism is complex, involving specific differentiation blockades and micro environmental conditions, highlighting the importance of epigenetic plasticity and stability throughout life. A variety of human disorders have been associated with epigenetic defects, including (but not limited to) neurodevelopmental syndromes, imprinting disorders, and complex diseases like cancer and diabetes. One of the most well-characterized examples of epigenetic inheritance is genomic imprinting, a process in which specific gene(s) are maintained in a parent-of-origin-specific manner, depending on whether they are inherited from the mother or the father. This selective monoallelic expression is essential for normal development, and its disruption can result in severe congenital disorders such as Prader-Willi [4, 5], Angelman [6, 7], and Beckwith-Wiedemann syndromes [8, 9].

Beyond imprinting, a growing body of research indicates that environmental factors – such as nutrition, stress, lifestyle, and toxin exposure – can induce epigenetic markers and gene expression changes in germ cells. These external stimuli can influence the way genes are expressed through epigenetic modifications and act as molecular switches that alter chromatin accessibility and gene expression. Such changes enable cells and tissues to adapt dynamically to external cues, which is especially critical during early development when the epigenome is most plastic. This feature can further be transferred across generations, resulting in a distinct epigenetic spectrum that influences disease susceptibility and phenotype. For example, maternal nutrition or exposure to endocrine-disrupting chemicals during pregnancy can program offspring metabolism and stress responses through persistent epigenetic marks. Importantly, certain environmentally induced epigenetic states are not fully erased during germ cell reprogramming, allowing them to be inherited across generations. This transgenerational epigenetic inheritance has been observed in both animals and humans, linking ancestral experiences – such as famine, trauma, or toxin exposure – to altered disease risk and physiological traits in descendants. Thus, epigenetics represents a crucial interface between gene alteration and environmental stimuli, explaining how external factors shape not only individual health and behavior but also exert long-lasting effects on evolutionary fitness and population health. These findings shed light on the developmental roots of health and disease, as well as their implications for public health and transgenerational risk assessment.

This chapter will explore the molecular underpinnings of epigenetic inheritance, with a focus on parent-of-origin-specific features, the epigenetic marks established and maintained, and the biological significance of these heritable regulatory states. Furthermore, we examine human disorders associated with disrupted epigenetic regulation and explore how recent advances in epigenomics are redefining concepts of heredity and evolutionary biology. Lastly, the emerging evidence of epigenetic modifiers as therapeutic strategies and their potential efficacy in the prevention and treatment of future clinical applications is also evaluated.

2. Epigenetic mechanisms involved in inheritance

Epigenetic inheritance, the transmission of gene regulatory information across generations without changes in the DNA sequence, can occur via distinct intergenerational and transgenerational mechanisms. Intergenerational inheritance refers to the direct passage of epigenetic marks from parents to their immediate offspring (F1 generation), where germline modifications persist through fertilization and influence gene expression in the developing embryo. Because F1 individuals develop within the parental environment, these effects may reflect both inherited epigenetic information and direct environmental exposure. In contrast, transgenerational inheritance involves the propagation of epigenetic information to the F2 generation or beyond, independent of continued exposure to the original environmental stimulus. This form of inheritance requires epigenetic marks to escape the extensive reprogramming events in germ cells and early embryos, including genome-wide DNA demethylation, resetting of histone modifications, and remodeling of non-coding RNA networks, thereby enabling a persistent epigenetic memory across multiple generations. Intergenerational effects are thus generally limited to F1 and influenced by the parental environment, whereas transgenerational effects are maintained in F2 or later generations through stable germline transmission. Examples include maternal stress impacting F1 offspring (intergenerational) versus environmental exposures in F0 inducing phenotypic changes in F2/F3 descendants (transgenerational). Distinguishing between these two forms of inheritance is critical for understanding heritable phenotypes, environmental influences, disease susceptibility, and the design of experiments to assess genuine epigenetic transmission.

Epigenetic inheritance is mediated by molecular mechanisms that modulate gene expression independently of changes in the DNA sequence, allowing gene expression patterns to be stably transmitted across cell divisions – and, in some cases, according to stage-, organ-, and cell-type-specific manners. During embryonic and post-embryonic development, gene expression patterns are precisely coordinated spatially and temporally to guarantee the proper establishment of cell identity and lineage specification, and finally, regulation of cell fate and differentiation. These processes are essential for early events like zygotic genome activation, gastrulation, organogenesis, and post-embryonic development. Meanwhile, epigenetic marks continue turning genes on or off at critical developmental stages to fine-tune gene expression in response to environmental signals, physiological needs, and tissue remodeling, thereby supporting growth, differentiation, and maintenance of tissue-specific functions.

Germline transmission of multiple layers of epigenetic marks involves the stable inheritance of epigenetic information – such as DNA methylation patterns, histone post-translational modifications, chromatin accessibility states, and regulatory non-coding RNAs – through the germline across generations. During gametogenesis and early embryonic development, extensive epigenetic reprogramming typically occurs and serves to reset the epigenome for totipotency. However, a subset of epigenetic marks can escape this reprogramming, particularly at imprinted loci, transposable elements, and certain regulatory regions, allowing for transgenerational epigenetic inheritance. These retained marks can influence chromatin architecture and transcriptional potential in the zygote and developing embryo. For example, histone modifications such as H3K27me3 or H3K4me3 may be preserved at specific loci and contribute to allele-specific gene expression. Additionally, small RNAs (e.g., piRNAs, endo-siRNAs) derived from the parental germline can guide epigenetic remodeling in the early embryo, reinforcing silencing of transposons or modulating gene expression. The coordinated transmission of these multiple epigenetic layers suggests a complex, multilayered regulatory system that integrates environmental cues, developmental signals, and genomic context to influence phenotypic outcomes across generations.

The most well-characterized mechanism is DNA methylation, typically occurring at CpG dinucleotides, which can silence gene expression and is faithfully copied during DNA replication by maintenance methyltransferases such as DNMT1. Histone modifications, including methylation, acetylation, and ubiquitination of histone tails, modulate chromatin structure and accessibility, contributing to transcriptional memory during development. These marks are established by writing enzymes (e.g., HATs, HMTs) and removed by erasers (e.g., HDACs, demethylases) and can be partially retained through cell division. Additionally, non-coding RNAs – including microRNAs, long non-coding RNAs (lncRNAs), and piRNAs – also play critical roles in directing chromatin, chromatin-modifying complexes to specific genomic regions, thereby reinforcing epigenetic states. Furthermore, chromatin architecture and three-dimensional genome organization contribute to the spatial

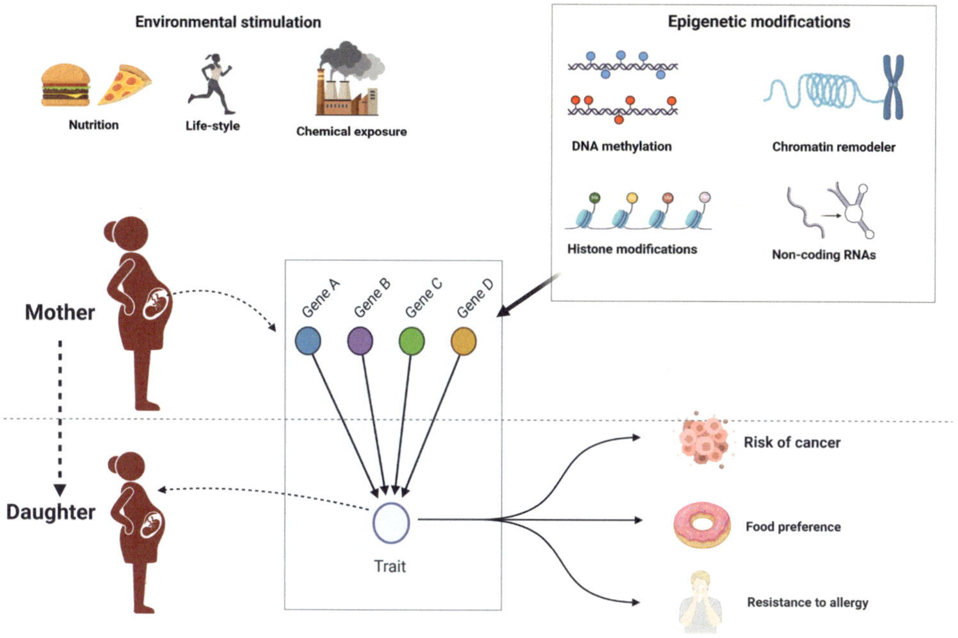

Figure 1.
Epigenetic modifications mediate intergenerational inheritance of traits. Environmental factors, including nutritional status, lifestyle choices, and exposure to toxicants, can induce epigenetic alterations in the mother and gene bodies. These modifications influence the transcriptional activity of multiple genes (genes A–D), leading to altered phenotypic traits. Some epigenetic marks can escape reprogramming during gametogenesis and early embryogenesis, thereby being transmitted to the offspring. As illustrated, such heritable modifications may contribute to increased disease susceptibility (e.g., cancer risk), changes in metabolic behavior (e.g., food preference), or enhanced resistance to allergy pathogens (e.g., pollen). This figure summarizes the role of epigenetic factors between environmental exposures and inherited biological outcomes.

maintenance of gene regulatory networks. Together, these interconnected mechanisms form a dynamic yet stable epigenetic landscape that governs cell identity and enables the transmission of regulatory information from parent to offspring (**Figure 1**).

2.1 DNA methylation

DNA methylation is one of the most well-established epigenetic mechanisms responsible for maintaining gene regulatory states across both somatic and germline cell divisions.

It primarily occurs at cytosine residues within CpG dinucleotides (the attachment of methyl groups to the C-5 position of the cytosine ring of DNA), where a methyl group is added by DNA methyltransferases (DNMTs). Among these enzymes, DNMT3A and DNMT3B are responsible for de novo methylation, while DNMT1 maintains existing methylation patterns during DNA replication to preserve epigenetic information through cell divisions.

In 2011, Shinsuke et al. identified that the oxidation of 5-methylcytosine (5mC) to 5-hydroxymethylcytosine (5hmC) in mammalian cells is catalyzed by ten-eleven translocation (TET) proteins, particularly TET1 [10, 11]. Tet-methylcytosine dioxygenase 2 (TET2) is responsible for removing the cytosine residues to maintain the DNA methylation-demethylation balance, further preserving the on and off status of gene expression. This is the first step of DNA demethylation, during which 5-hydroxymethylcytosine (5hmC) is further oxidized to 5-formylcytosine (5fC) and 5-carboxylcytosine (5caC), and subsequently restored to cytosine through a base excision repair pathway mediated by thymine DNA glycosylase (TDG).

Methylation of promoter regions generally reduces the accessibility of transcription factors to their DNA binding sites, which is usually associated with transcriptional repression, serving as a stable silencing mechanism for specific genes. Methylation of promoter regions may also recruit histone modification complexes to further modify histones, such as H3K9me3 methyltransferase, ultimately silencing gene transcription. However, promoter methylation does not invariably lead to gene silencing; it is context-specific and depends on the binding proteins, either transcription activator or repressor, and the recruitment of other modification proteins (e.g., NRF1, MeCP2, HDACs). Mendoza et al. used an artificial epigenome editing technique to test how induced DNA methylation affects endogenous promoter regulation in MCF-7 cells. They found that the simultaneous methylation of promoters resulted in undetectable repression of gene expression; it was heavily dependent on TF binding affinities and context-specific epigenomic features [12]. DNA methylation exhibits a greater level of complexity than previously recognized [13]. Recent studies have also reported inverse correlations between DNA methylation and gene expression during the onset and progression of various cancer types [14]. Collectively, these findings suggest that the relationship between DNA methylation and gene expression is context-dependent, varying with both cell type and cell cycle phase. A healthy cell cycle requires a moderate level of methylcytosine (5mC) in the whole genome, and the methyltransferase families and TET families are responsible for methylation maintenance [15, 16].

In the context of epigenetic inheritance, DNA methylation is extensively established across different cell types and plays key roles in genomic imprinting, X-chromosome inactivation, the stable silencing of specific genes [17–19], and age-related phenotypic changes. DNA methylation patterns can be faithfully propagated during mitosis, ensuring that daughter cells retain their epigenetic identity. Remarkably, a subset of these methylation marks – particularly those associated with genomic imprinting – can also evade the global epigenetic reprogramming that typically occurs during gametogenesis and early embryonic development. Parent-of-origin-specific methylation marks are established in the germline and preserved through fertilization, resulting in monoallelic expression of imprinted genes in offspring. Disruption of these methylation patterns can cause developmental disorders and has been associated with transgenerational effects of environmental exposures, underscoring the critical role of DNA methylation in epigenetic inheritance.

For instance, studies in humans have identified specific methylation signatures associated with prenatal exposures (e.g., maternal diet, stress, or environmental toxins) that may influence the health and disease risk of the next generation. Studies on the Dutch Hunger Winter (1944–1945) revealed that individuals exposed to famine in utero exhibited altered methylation at the IGF2 (Insulin-like Growth Factor 2) gene even decades later [20]. These epigenetic changes were associated with an increased risk of metabolic disorders, including obesity and type 2 diabetes.

Similarly, maternal stress during pregnancy has been implicated in altered methylation of genes regulating the hypothalamic-pituitary-adrenal (HPA) axis, such as NR3C1, which encodes the glucocorticoid receptor and influences stress reactivity and mental health outcomes in offspring [21, 22]. In terms of environmental toxins, prenatal exposure to substances like bisphenol A (BPA) or air pollutants has been shown to cause differential methylation levels of genes involved in immune response, neural development, and cancer susceptibility [23, 24]. These methylation patterns can persist after birth and influence disease risk throughout life.

Fraga et al. were the first to report age-related changes in DNA methylation patterns, highlighting the dynamic nature of the epigenome over the human lifespan [25]. The genetically identical monozygotic twins are not identical; they had different risks of disorders, such as cardiovascular diseases, neurodegenerative disorders, and cancer. Despite sharing similar diets, lifestyles, and environmental factors, distinct profiles of DNA methylation and histone acetylation across different cell types throughout the lifespan may underlie their differential susceptibility and onset of common diseases. Actually, DNA methylation changes did correlate with chronological age within human tissues and cell types, indicating that "methylation clocks" may be capable of correlating with developmental conditions, including cancer [26], neurodegeneration [27, 28], and aging-related chronic diseases [29]. A clear correlation between aging and DNA methylation has been observed across multiple mouse tissues, with genes implicated in aging – such as EDARADD, TOM1L1, and NPTX2 – frequently exhibiting hypermethylation within their promoter CpG islands [30–32]. In addition, DNA methylation has been employed as an "epigenetic clock" to estimate chronological age and has been proven to be a more accurate predictor of an individual's health status and susceptibility to age-related diseases [33–35]. Bocklandt et al. were the first to report age-associated changes in DNA methylation and their contribution to age-related diseases, based on the analysis of saliva samples from 34 pairs of identical twins aged 21–55 years [32]. Subsequently, Hannum et al. developed a highly accurate epigenetic clock by constructing a quantitative aging model based on DNA methylation measurements at more than 450,000 CpG markers in whole blood samples from 656 individuals aged 19–101 years, highlighting the specific components of the aging process with the profile of methylation in age-related disease [36]. Recent genome-wide methylation analyses have shown that age-related CpG methylation changes occur independently of tissue type, gender, and cell type. Building on this, Horvath et al. (2018) developed a novel and highly robust DNA methylation (DNAm) age estimator applicable to a wide range of human cell types – including fibroblasts, keratinocytes, buccal cells, endothelial cells, and lymphoblastoid cells – as well as tissues such as skin, blood, and saliva, enabling accurate tracking of the dynamic aging process in *ex vivo* human cell aging assays [37].

Though much remains to be clarified about how stable these methylation marks are across generations, animal studies suggest that some epigenetic modifications can escape reprogramming during gametogenesis and early embryogenesis, allowing for transgenerational epigenetic inheritance. In humans, such inheritance is harder to prove directly, but accumulating evidence suggests that DNA methylation may mediate the biological memory of environmental conditions experienced by parents, thereby shaping the health trajectories of their children and grandchildren.

2.2 Histone modifications

Nuclear DNA is primarily packaged around histone proteins to form chromatin, which is further organized into chromosomes [38–40]. Histones form a stable and robust structure to strongly support the flexible and fragile DNA. The unit of histones may undergo depolymerization and reorganization before and after the formation of DNA replication and transcription complexes [41, 42]. In the chromosome, histones are constructed into octamers, and then the rope-like DNA winds around each octamer in 1.67 circles, leaving some histone tails outside [43]. Since DNA is buried in the histone grooves during the transcription process, the RNA Pol II-containing machinery has to search for and localize the gene promoter region. Therefore, histones serve as a "gatekeeper" to either facilitate or suppress DNA transcription [44]. Histone modification can occur in nearly the entire amino acid sequence, and most often, the modification process is centered around the histone tails, particularly at the N-terminals drifting outside the histone octamer core, although sometimes it can also occur inside the octamer [45, 46]. These modifications consist of acetylation, methylation, ubiquitination, phosphorylation, sumoylation, etc. The modifications to histone tails function as a communication mechanism with either the transcription apparatus or other proteins, signaling "on" or "off" for gene transcription. Through this modification, these histone tails can also recruit other functional complexes to regulate gene expression together.

Among histone modifications, methylation and acetylation are the most extensively studied due to their critical roles in regulating chromatin structure and gene expression, and these two modifications usually target lysine concurrently and sometimes alternately. To control gene expression, both the inhibitory and stimulating factors should be maintained at optimal levels. Acetylation is generally associated with gene activation, whereas other histone modifications can be linked to either transcriptional activation or repression, depending on the specific modification and context. For instance, histone modifications such as H3K4me3 and H3K36me3 enhance the accessibility of the transcription machinery to gene promoter regions, particularly at transcription start sites (TSSs), thereby promoting gene expression; on the other hand, modifications at H3K9me3, H3K27me3, and H4K20me3 can result in a down-regulation or shut down of gene transcription [47]. Histone modifications often interact with DNA methyltransferases and the chromatin remodeling complex, and some of the histone modification markers can also recruit antagonistic complexes [48], all of which work synergistically to guarantee more accurate and flexible gene expression.

Histone modifications encompass a variety of covalent alterations, including methylation/demethylation, acetylation/deacetylation, phosphorylation/dephosphorylation, and ubiquitination/deubiquitination, each of which plays a critical role across cell divisions and, in some cases, across generations. In humans, these chemical tags on histone proteins help determine which genes are active or silent, contributing to cell identity, development, and disease susceptibility. This process ensures the continuity of chromatin states and preserves cell identity. While most histone marks are erased during the phase of extensive epigenetic reprogramming that occurs in gametes and early embryos, a subset – particularly those associated with heterochromatin and imprinting – can escape erasure. The retention of specific histone modifications enables certain chromatin marks to influence gene expression in offspring, thereby contributing to transgenerational epigenetic inheritance. For instance, paternal lifestyle factors like diet, stress, or exposure to toxins have been shown to influence histone marks in sperm, which may, in turn, affect gene expression and development in the offspring. Although the mechanisms in humans are still not fully understood, histone-based inheritance is a growing area of interest, with implications for understanding how environmental experiences can shape biology beyond DNA sequence alone. This insight highlights the importance of histone dynamics not only in immediate gene regulation but also in transgenerational health outcomes.

2.3 Non-coding RNAs

Non-coding RNAs (ncRNAs) have emerged as important regulators of epigenetic modification and gene regulation. Transcribed from DNA but not translated into proteins, these RNAs play diverse roles in chromatin remodeling, transcriptional regulation, and post-transcriptional control. The most extensively studied classes of ncRNAs include microRNAs (miRNAs), short interfering RNAs (siRNAs), PIWI-interacting RNAs (piRNAs), and long non-coding RNAs (lncRNAs) [49–51].

Additional types of non-coding RNAs, such as circular RNAs (circRNAs), are also emerging as important regulators of diverse cellular processes. Although the ncRNA family is vast – comprising thousands of molecules – some members are thought to be non-functional, historically labeled as "junk RNAs" [52]. However, growing evidence shows that ncRNAs play essential roles in controlling gene expression and epigenetic regulation [53–56]. One major mechanism is RNA interference (RNAi), a pathway in which ncRNAs regulate gene expression post-transcriptionally. In this process, double-stranded RNAs (dsRNAs) are cleaved by Dicer into short siRNAs, which are then incorporated into the RNA-induced silencing complex (RISC). Guided by siRNAs, Argonaute proteins within RISC identify complementary mRNAs and degrade them, effectively silencing translation [57].

Non-coding RNAs are now recognized as central players in epigenetic inheritance, acting as molecular carriers of information that can transmit regulatory states across generations. Unlike coding RNAs, ncRNAs influence gene expression without encoding proteins. They regulate chromatin structure, DNA methylation, and histone modifications, thereby shaping how genes are expressed in offspring. For example, sperm- and oocyte-derived small RNAs can escape epigenetic reprogramming during fertilization and deliver regulatory cues that affect embryonic development and even long-term traits. Through mechanisms like RNA interference, chromatin remodeling, and interaction with Argonaute proteins, ncRNAs help determine which genetic programs are activated or silenced. Their ability to preserve or alter epigenetic marks highlights their role as key regulators of transgenerational inheritance, linking parental experiences or environmental exposures to phenotypic outcomes in descendants.

2.4 Chromatin remodeling and higher-order structure

Genetic information is stored and protected in the form of chromatin, composed of highly wrapped DNA and histones; during DNA replication, transcription, or repair processes, the tightly packaged DNA needs to be loosened at specific positions to become accessible to various enzymes. Chromatin during gene transcription functions like a spring, adopting a more or less condensed state according to the molecular events happening in the nucleus. Chromatin remodeling is necessary for changing the density of the nucleosomes, which loosens the chromatin, facilitates interaction with transcription factors, and promotes the processes of DNA replication, transcription, and repair.

Determining a gene's expression status, and the degree to which it is silenced or activated, depends on multiple factors, including the epigenetic state of the gene, its chromatin context, and the configuration of promoter and enhancer regions. Chromatin remodelers, which are ATP-dependent DNA translocases, generate the force necessary to reposition nucleosomes, thereby dynamically regulating chromatin structure. Importantly, nucleosome positions and chromatin states are not static but continuously change, allowing fine-tuned control of gene expression. The plant homeodomain- and bromodomain-containing chromatin remodelers can recognize histone modifications and the chromatin architecture by altering nucleosome positions. Meanwhile, certain histone modification markers can recruit specific chromatin remodeling complexes, thereby reinforcing nucleosome repositioning and chromatin remodeling. During this process, chromatin remodelers play a critical role in mediating the "position effect" on gene expression, where the spatial arrangement of nucleosomes and chromatin structure influences the accessibility of transcriptional machinery and ultimately regulates gene activity. Nucleosome positioning is not random; instead, it is tightly regulated to either condense or relax chromatin at specific genomic loci. In human cells, chromatin remodelers are organized into four major families – SWI/SNF, ISWI, CHD, and INO80 – each of which utilizes ATP-dependent mechanisms to modulate nucleosome positioning and chromatin accessibility, thereby influencing gene expression. By modulating chromatin structure to regulate gene expression, chromatin remodeling complexes are essential for the normal growth, development, and cellular differentiation of multicellular organisms. Aberrant chromatin remodeling, resulting from dysfunctional remodelers, is frequently associated with neurodevelopmental disorders and intellectual disabilities, and can also contribute to immunodeficiency and muscle-wasting syndromes.

Chromatin remodelers are essential regulators of epigenetic inheritance because they govern DNA accessibility and maintain the stability of epigenetic marks that can be transmitted across generations. These ATP-dependent complexes restructure nucleosomes by sliding, ejecting, or replacing histones, thereby influencing which genomic regions remain transcriptionally active or repressed. During gametogenesis and early embryonic development, chromatin remodelers work with DNA methyltransferases and histone-modifying enzymes to erase most parental epigenetic marks while selectively preserving imprinted regions and other loci that escape reprogramming. By maintaining or re-establishing specific chromatin states, chromatin remodelers help safeguard the transmission of essential regulatory information from parents to offspring. Their activity ensures that developmental programs are reset for a "blank slate," yet still permit selective inheritance of epigenetic memory, linking parental environment or physiology to gene expression patterns in the next generation.

3. Parent-of-origin maintenance and epigenetic plasticity

Parent-of-origin maintenance and epigenetic plasticity represent two complementary forces that shape gene regulation across development and disease.

Parent-of-origin maintenance is most clearly illustrated by genomic imprinting, where epigenetic modifications mark certain alleles based on whether they are inherited maternally or paternally. Unlike typical Mendelian inheritance, where both maternal and paternal alleles are equally active, imprinted genes are epigenetically marked so that only one allele – either maternal or paternal – is expressed, while the other is silenced. These imprints are established in the germline, survive post-fertilization reprogramming, and must be preserved through subsequent cell divisions to maintain proper parent-specific expression. Disruption of this process underlies imprinting disorders, where the loss of paternal or maternal allele function, respectively, leads to developmental and neurological defects. This highlights the necessity of faithful epigenetic inheritance for dosage-sensitive genes.

Cellular plasticity denotes the intrinsic ability of cells to undergo dynamic and reversible changes in identity, phenotype, and functional state in response to both intrinsic genetic programs and extrinsic environmental cues. This adaptability is orchestrated by tightly regulated epigenetic mechanisms that collectively shape the transcriptional landscape and contribute to the maintenance or alteration of lineage fidelity. Under physiological conditions, plasticity is essential for embryonic development, tissue regeneration, and homeostatic maintenance. However, in pathological contexts such as cancer, (pre)malignant cells can aberrantly co-opt these plasticity programs to facilitate malignant transformation, tumor initiation, progression, and resistance to therapy. Elevated plasticity enables these cells to circumvent fixed lineage constraints, adopt stem-like or mesenchymal states, and dynamically rewire their gene expression profiles in response to microenvironmental pressures – including hypoxia, inflammation, and therapeutic stress. This phenotypic fluidity contributes to intratumoral heterogeneity, immune evasion, metastatic competence, and the emergence of drug-tolerant persister populations. Deciphering the epigenetic circuits that regulate cellular plasticity and their dysregulation in cancer may uncover novel therapeutic targets aimed at re-establishing lineage stability or preventing adaptive resistance.

By contrast, epigenetic plasticity emphasizes the system's capacity to remodel these marks in response to developmental cues or environmental pressures. During early embryogenesis, global demethylation resets most of the genome, allowing pluripotent cells to regain broad developmental potential before lineage-specific methylation patterns are established. Similarly, induced pluripotent stem cell (iPSC) reprogramming exploits this plasticity: transcription factors erase somatic epigenetic landscapes and re-establish a pluripotent state. At the organismal level, environmental factors – including diet, stress, or toxin exposure – can induce alterations in DNA methylation or chromatin state, thereby reshaping gene expression programs in ways that may persist across cell divisions.

Genomic imprinting plays essential roles in embryonic growth, placental function, and neurodevelopment, and errors in imprinting can result in imprinting disorders, congenital overgrowth conditions, or even contribute to cancers. Additionally, research suggests that environmental inputs can influence imprinting and potentially affect the health and disease risk of future generations, making genomic imprinting a key mechanism linking environment, gene regulation, and inherited phenotypes. These imprints are established during the formation of gametes (sperm or egg) through DNA methylation and histone modifications, and they are maintained throughout development. Although the DNA sequence itself

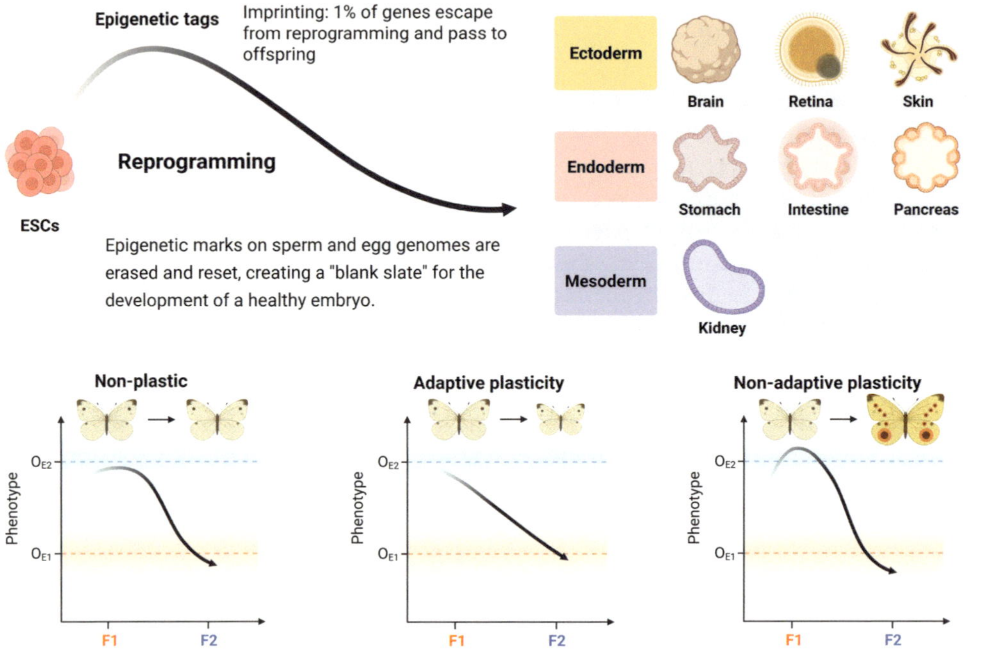

Figure 2.
Inherited phenotypic maintenance by epigenetic plasticity. Epigenetic reprogramming in early embryonic stem cells (ESCs) erases most parental epigenetic marks on sperm and egg genomes, generating a "blank slate" for development. However, ~1% of genes retain parental imprinting marks, allowing certain epigenetic information to pass to the offspring. These imprints influence the differentiation of the three germ layers – ectoderm (brain, retina, skin), endoderm (stomach, intestine, pancreas), and mesoderm (kidney) – which give rise to diverse tissues. The lower panels illustrate three models of transgenerational phenotypic inheritance. In the non-plastic model, the phenotype is fixed and does not adjust across environments, leading to reduced fitness in subsequent generations (F1, F2). In adaptive plasticity, phenotypes shift gradually toward the optimal environment across generations, maintaining alignment with environmental demands. In contrast, non-adaptive plasticity produces maladaptive phenotypic shifts, moving away from optimal conditions and reducing offspring fitness.

remains unchanged, the epigenetic marks are heritable, allowing certain expression patterns to be passed to the next generation (**Figure 2**).

4. Epigenetic-mediated brain development during inheritance

There are thousands of expansive genomic loci and distal regulatory elements in the germ cells, with a variety of epigenetic markers during production. Then, reprogramming erased most of these epigenetic tags, including genome-wide loss of DNA methylation, localized hypermethylation of CpG promoters of specific genes, broad changes in histone modifications, alterations in chromatin structure, and deregulation of the ncRNA network and other molecular targets.

During brain development, the finely tuned balance of epigenetic modifications regulates gene expression programs critical for neurogenesis, neuronal differentiation, and maturation. These processes involve dynamic and coordinated changes in DNA methylation, histone modifications (e.g., H3K4me3, H3K27me3), chromatin accessibility, and non-coding RNA activity, which collectively shape the transcriptional landscape of neural progenitor cells and their progeny. During early embryonic development, epigenetic reprogramming establishes lineage-specific gene expression patterns that guide the specification of neural stem cells and their subsequent differentiation into diverse neuronal and glial subtypes. Maintenance of epigenetic homeostasis ensures proper timing and spatial organization of neurodevelopmental events, such as cortical layering, synaptogenesis, and circuit formation.

Disruption of this balance – whether through genetic mutations in epigenetic regulators (e.g., DNMTs, TETs, HDACs, or Polycomb group proteins), environmental insults, or metabolic stress – can lead to aberrant gene expression, impaired neuronal identity, and increased susceptibility to neurodevelopmental disorders, schizophrenia, and intellectual disability. In addition, altered epigenetic patterns have been observed across multiple brain cell types, including microglia, astrocytes, and neurons, highlighting the widespread impact of epigenetic regulation on neural function.

Maternal drug or alcohol use during pregnancy is a major risk factor for hypoxic-ischemic encephalopathy (HIE), a severe birth complication that can affect full-term infants. HIE affects an estimate of 1–6 infants out of every 1,000 births. In circumstances of hypoxia, microglia are activated and undergo various epigenetic changes that influence their behavior in clearing debris and initiating repair processes, including their inflammatory response, phagocytic activity, and neuroprotective functions. Astrocytes, characterized as star-shaped glial cells in the brain, play crucial roles in maintaining brain homeostasis, supporting neurons, and responding to injury. After injury, astrocytes become reactive (a process called reactive astrogliosis) and undergo several epigenetic changes that impact their behavior. These changes influence astrocyte functions, such as neuroinflammation, tissue repair, and neuroprotection. After repair, neuroplasticity experiences significant epigenetic changes that impact survival, function, and the ability to recover. These changes affect synaptic plasticity, axonal growth, and neural regeneration, contributing to functional recovery and rehabilitation. Peripheral infiltrated cells, which include various immune cells from the periphery, such as macrophages, neutrophils, T cells, and B cells, experience significant epigenetic changes as they interact with injured brain tissue. These changes regulate their activation, differentiation, and function, influencing both the acute inflammatory response and long-term recovery processes. These epigenetic alterations modulate the genes involved in inflammation, immune cell polarization, oxidative stress, and apoptosis, thereby influencing both the progression of brain injury and the potential for recovery. Understanding these epigenetic mechanisms can open new therapeutic avenues aimed at modulating immune cell responses to promote neuroprotection and recovery.

Epigenetic and genomic imprinting-related diseases often arise from abnormal DNA methylation patterns, mutations in imprinting control regions (ICRs), or uniparental disomy (inheritance of both copies of a chromosome from one parent). For example, Prader-Willi syndrome (PWS) and Angelman syndrome (AS) both involve the 15q11-q13 region but differ in the parental allele affected – PWS results from loss of paternal gene expression, while AS results from loss of maternal UBE3A expression. Similarly, Beckwith-Wiedemann syndrome (BWS) and Silver-Russell syndrome (SRS) involve dysregulation of imprinted genes on chromosome 11p15, affecting growth-promoting genes like IGF2 and H19 (**Table 1**). These disorders underscore the critical role of epigenetic homeostasis in development, as even subtle disruptions in imprinting can lead to profound phenotypic consequences, including growth abnormalities, neurodevelopmental delays, and increased cancer risk.

Disease	Epigenetic mechanism	Description
Prader-Willi syndrome	Genomic imprinting	Caused by a lack of paternal gene expression on chromosome 15q11-q13.
Angelman syndrome	Genomic imprinting	Results from maternal gene loss or silencing occur on the same chromosomal region.
Beckwith-Wiedemann syndrome	Genomic imprinting	Overgrowth disorder linked to alter imprinting at chromosome 11p15.
Silver-Russell syndrome	Genomic imprinting	Growth restriction associated with hypomethylation on chromosome 11.
Fragile X syndrome	DNA methylation	Expansion of CGG repeats leads to the silencing of the FMR1 gene.
Rett syndrome	MECP2 mutation affects epigenetic regulation	X-linked disorder causes severe neurodevelopmental regression.

Table 1.
Epigenetic and genomic imprinting-related diseases.

Understanding the molecular basis of these diseases not only provides insight into developmental biology but also opens avenues for epigenetic therapies aimed at restoring normal imprinting patterns.

5. Targeting epigenetic modifiers to overcome inherited diseases

Epigenetic modifications work synergistically to maintain the stability of gene transcription; however, dysregulation of epigenetic modifications can lead to aberrant gene expression and contribute to disease. While the maintenance of epigenetic homeostasis is essential for normal cellular development, perturbations in these regulatory mechanisms are also implicated in a variety of pathological conditions. In fact, epigenetic modifications have been found to contribute to aging [58], the occurrence of cancer, autoimmune diseases [59, 60], heart disease, mental illness, etc. [61]. Given that epigenetic modifications are inherently reversible, they represent attractive targets for therapeutic intervention. Current epigenetic treatments aim to modulate DNA methylation or histone acetylation patterns. Furthermore, changes in epigenetic modifications hold considerable promise as early diagnostic indicators, potentially allowing detection of disease before clinical onset. In cancer, besides point mutations or chromatin depletion, there are numerous mutations in epigenetic modifications compared to normal healthy cells. Numerous aberrations in epigenetic modifications are observed compared with normal healthy cells. These early-occurring epigenetic "trademarks" contribute to the maintenance of cancer cell properties, including self-renewal and unlimited proliferative capacity [62, 63].

Epigenetic modifier drugs are therapeutic agents designed to target enzymes and pathways that regulate epigenetic modifications. Several drugs with notable epigenetic activity, including azacitidine, decitabine, and various histone deacetylase inhibitors (HDACi), have shown promising effects in treating inherited diseases caused by epigenetic dysregulation, such as imprinting disorders and certain monogenic syndromes (**Table 2**). For example, DNA methyltransferase inhibitors (e.g.,

Drug/therapy	Class/mechanism	Inherited disease	Clinical status	Notes
Nusinersen (Spinraza)	Antisense oligonucleotide (ASO) modifies SMN2 pre-mRNA splicing to increase SMN protein.	Spinal muscular atrophy (SMA)	FDA-/EMA-approved	First FDA-approved RNA-based therapy, delivered intrathecally.
Onasemnogene abeparvovec (Zolgensma)	Gene therapy: AAV vector delivers functional SMN1 gene (affects gene expression via epigenetic regulation of splicing/repair).	SMA	FDA-/EMA-approved	One-time systemic infusion; complements ASO therapy.
Eteplirsen (Exondys 51)	ASO: exon-skipping therapy restores dystrophin reading frame.	Duchenne muscular dystrophy (DMD)	FDA-accelerated approval	Targets specific exon mutations; several similar ASOs in the pipeline.
Valproic Acid	HDACi increases histone acetylation, activating gene expression.	Epilepsy, bipolar disorder (not an inherited disease-specific condition), and trials in Friedreich's Ataxia	FDA-approved (for epilepsy); repurposed in clinical trials for inherited ataxias	Mixed results; safety concerns at high doses.
Mipomersen (Kynamro)	ASO inhibits ApoB-100 mRNA to reduce LDL.	Familial hypercholesterolemia (FH)	FDA-approved (restricted use)	Demonstrates ASO potential in metabolically inherited disorders.

Table 2.
Epigenetic drugs that are approved by the FDA for inherited diseases.

azacitidine and decitabine) can reverse aberrant gene silencing in conditions like Fragile X syndrome, where hypermethylation of the FMR1 gene leads to intellectual disability. Similarly, HDACi compounds like valproic acid and vorinostat can modulate chromatin structure to restore gene expression in disorders like Rett syndrome, which is caused by mutations in the MECP2 gene, a critical epigenetic reader. In imprinting disorders such as Angelman syndrome, emerging strategies aim to silence the paternal UBE3A-antisense transcript using antisense oligonucleotides, thereby reactivating the silenced maternal allele. These approaches highlight the therapeutic potential of epigenetic drugs in correcting gene dosage imbalances and restoring normal gene function in diseases rooted in heritable epigenetic errors. As our understanding of epigenetic regulation deepens, these drugs may offer targeted, reversible, and personalized treatments for a growing spectrum of inherited disorders.

6. Challenges and future perspectives

Epigenetic information in germ cells is inherently complex, highly dynamic, and challenging to measure and quantify due to its multilayered nature and sensitivity to environmental and developmental cues. Germ cells experience extensive epigenetic reprogramming during gametogenesis and early embryogenesis, characterized by global DNA demethylation and the re-establishment of sex-specific imprinting marks. However, not all epigenetic marks are erased – some, such as imprinted loci and transposable element silencing, are retained and transmitted across generations. The variability of epigenetic states in germ cells, influenced by environmental factors, can lead to inter-individual differences in epigenetic profiles. Moreover, the technical limitations of current epigenomic tools – such as low input requirements, single-cell resolution, and distinguishing between active and poised chromatin states – make it difficult to comprehensively map and interpret epigenetic landscapes in germ cells. This complexity poses significant challenges for understanding how epigenetic information contributes to heredity, development, and disease susceptibility.

Environmental exposure-induced epigenetic changes in the germline represent a compelling mechanism by which ancestral experiences can influence the health and development of subsequent generations. Despite growing evidence, establishing causal links between environmental exposures and heritable epigenetic changes remains challenging. This is due to the complexity of the epigenome, its dynamic nature, and the difficulty in isolating germline-specific effects from confounding somatic or maternal influences. Recent advances in single-cell multiomics, sperm RNA profiling, and epigenome editing have begun to provide mechanistic insights and causal evidence. For example, studies have shown that specific classes of sperm RNAs, such as tRNA-derived fragments (tRFs) and piRNAs, can mediate the inheritance of environmentally induced phenotypes [64]. Moreover, experiments using controlled exposure windows and lineage tracing have helped distinguish true transgenerational inheritance (effects seen in the F3 generation and beyond) from multigenerational exposure (effects in directly exposed F1 and F2 generations) [65, 66]. These findings underscore the importance of rigorous experimental design and interdisciplinary approaches to unravel the molecular basis of germline epigenetic inheritance. As the field progresses, it holds promise for identifying biomarkers of exposure, understanding disease etiology, and informing public health strategies aimed at mitigating long-term risks from environmental insults.

Recent advances in technology have transformed the study of epigenetic regulation and inheritance, providing unprecedented resolution and control over chromatin states. Engineered artificial epigenome editing enables precise manipulation of DNA methylation or histone modifications at specific genomic loci, allowing researchers to dissect causal relationships between epigenetic marks and gene expression, and offering potential therapeutic strategies for correcting aberrant epigenetic states in inherited or acquired diseases. Highly accurate DNA methylation clocks provide quantitative measures of biological age and epigenetic drift, facilitating early detection of age-related and epigenetically mediated disorders. Additionally, multi-omics approaches, which integrate genomics, epigenomics, transcriptomics, and proteomics, allow comprehensive mapping of the regulatory networks that coordinate gene expression and epigenetic inheritance. Collectively, these technologies not only enhance mechanistic insights into epigenetic dynamics but also pave the way for novel diagnostic tools and targeted interventions in diseases driven by epigenetic dysregulation.

Future research should focus on: (1) elucidating the mechanisms underlying the interactions between environmental factors and the resulting epigenetic alterations; (2) developing (multichannel) targeted therapies that address multiple aspects of epigenetic dysfunctions; (3) investigating the long-term accumulation effects of environmental stimulators across generations; (4) exploring the role of epigenome editors (targeted methylation/demethylation or histone modifiers) and their potential as therapeutic agents; and (5) studying the impact of aging and strategies to mitigate its effects. By addressing these areas, we can improve our understanding of epigenetics involved in cell physiology and develop more effective interventions for inherited disorders, tumors, and age-related neurodegenerative diseases.

7. Concluding remarks

In spite of decades of studies on epigenetic aberration and inherited diseases, the underlying mechanisms are still not fully understood. Epigenetic transgenerational inheritance within animal models cannot fully mimic the complexity of humans. In this chapter, we highlighted epigenetic changes, especially the epigenetic-related gene features in transgenerational inheritance. We then described the commonly studied epigenetic markers and discussed how these epigenetic markers are transferred across generations. Finally, we addressed strategies for modulating innate epigenetic regulators through pharmacological agents in the context of developing novel therapies for inheritance-related diseases. Recent FDA-approved epigenetic drugs and their target diseases are also discussed. This chapter aims to provide researchers with an overview of recent advances in epigenetic mechanisms involved in transgenerational inheritance. The factors highlighted here may inform the development of innovative treatment strategies and potential therapies for epigenetically mediated inherited disorders.

Author contributions

S.Q. had the initial idea for this chapter; S.Q. and J.D. designed the study; J.D. and A.R. collected literature and prepared the table; A.R. and S.Q. performed the figure preparation; J.D., S.Q., and A.R. wrote the manuscript; S.Q. and A.R. proofread the whole manuscript; and all authors approved the final manuscript.

Conflict of Interest

The authors declare that the research was conducted in the absence of any commercial or financial relationships that could be construed as a potential conflict of interest.

Author details

Jiabin Duan[1], Md Abdur Razzaque[2] and Shaohua Qi[3]*

1 Department of Physiology, University of Texas Southwestern Medical Center, Dallas, USA

2 Department of Medicine–Nephrology and Hypertension, Tulane University School of Medicine, New Orleans, USA

3 Department of Neurology, McGovern Medical School, The University of Texas Health Science Center at Houston, Houston, USA

*Address all correspondence to: qsh7950@gmail.com

References

[1] de Groote ML, Verschure PJ, Rots MG. Epigenetic Editing: Targeted rewriting of epigenetic marks to modulate expression of selected target genes. Nucleic Acids Research. 2012;**40**(21):10596–10613

[2] Kim M, Costello J. DNA methylation: An epigenetic mark of cellular memory. Experimental & Molecular Medicine. 2017;**49**(4):e322–e322

[3] Center GSL. Epigenetics & Inheritance. 2013, July 15. Available from: https://learn.genetics.utah.edu/content/epigenetics/inheritance/ [Accessed: 2025-August-27]

[4] Mendiola AJP, LaSalle JM. Epigenetics in Prader-Willi Syndrome. Frontiers in Genetics. 2021;**12**:624581

[5] Salles J, et al. Patients with PWS and related syndromes display differentially methylated regions involved in neurodevelopmental and nutritional trajectory. Clinical Epigenetics. 2021;**13**:159

[6] Lalande M, Calciano MA. Molecular epigenetics of Angelman syndrome. Cellular and Molecular Life Sciences. 2007;**64**(7-8):947–960

[7] Chamberlain SJ, Lalande M. Angelman syndrome, a genomic imprinting disorder of the brain. Journal of Neuroscience. 2010;**30**(30):9958–9963

[8] Soejima H, Higashimoto K. Epigenetic and genetic alterations of the imprinting disorder Beckwith–Wiedemann syndrome and related disorders. Journal of Human Genetics. 2013;**58**(7):402–409

[9] Papulino C, et al. Preclinical and clinical epigenetic-based reconsideration of Beckwith-Wiedemann Syndrome. Frontiers in Genetics. 2020;**11**

[10] Ito S, et al. Role of Tet proteins in 5mC to 5hmC conversion, ES-cell self-renewal and inner cell mass specification. Nature. 2010;**466**:1129–1133

[11] Dawlaty MM, et al. Tet1 is dispensable for maintaining pluripotency and its loss is compatible with embryonic and postnatal development. Cell Stem Cell. 2011;**9**:166–175

[12] de Mendoza A, et al. Large-scale manipulation of promoter DNA methylation reveals context-specific transcriptional responses and stability. Genome Biology. 2022;**23**:163

[13] Smith J, et al. Promoter DNA Hypermethylation and Paradoxical Gene Activation. Trends in Cancer. 2020;**6**:392–406

[14] Spainhour JC, et al. Correlation patterns between DNA methylation and gene expression in The Cancer Genome Atlas. Cancer Informatics. 2019;**18**:1176935119828776

[15] Vilkaitis G, et al. Processive methylation of hemimethylated CpG sites by mouse Dnmt1 DNA methyltransferase. Journal of Biological Chemistry. 2005;**280**:64–72

[16] Araujo FD, et al. The DNMT1 Target Recognition Domain resides in the N Terminus. Journal of Biological Chemistry. 2001;**276**:6930–6936

[17] Wolf SF, Migeon BR. Studies of X chromosome DNA methylation in normal human cells. Nature. 1982;**295** (5851):667–671

[18] Singer-Sam J, Riggs AD. X chromosome inactivation and DNA methylation. In: Jost J-P, Saluz H-P, editors. DNA Methylation: Molecular Biology and Biological Significance. Birkhäuser Basel: Basel; 1993. p. 358–384

[19] Sharp AJ, et al. DNA methylation profiles of human active and inactive X chromosomes. Genome Research. 2011;**21**:1592–1600

[20] Heijmans BT, et al. Persistent epigenetic differences associated with prenatal exposure to famine in humans. Proceedings of the National Academy of Sciences of the United States of America. 2008;**105**:17046–17049

[21] Chalfun G, et al. Perinatal stress and methylation of the NR3C1 gene in newborns: Systematic review. Epigenetics. 2022;**17**:1003–1019

[22] Nazzari S, et al. Maternal and infant NR3C1 and SLC6A4 epigenetic signatures of the COVID-19 pandemic lockdown: When timing matters. Translational Psychiatry. 2022;**12**:386

[23] Perera F, Herbstman J. Prenatal environmental exposures, epigenetics, and disease. Reproductive Toxicology. 2011;**31**(3):363–373

[24] Klibaner-Schiff E, et al. Environmental exposures influence multigenerational epigenetic transmission. Clinical Epigenetics. 2024;**16**:145

[25] Fraga MF, et al. Epigenetic differences arise during the lifetime of monozygotic twins. Proceedings of the National Academy of Sciences. 2005;**102**:10604–10609

[26] Fernandes Durso D, et al. Acceleration of leukocytes' epigenetic age as an early tumor and sex-specific marker of breast and colorectal cancer. Aging. 2017;**8**(14):23237–23245

[27] Horvath S, Ritz BR. Increased epigenetic age and granulocyte counts in the blood of Parkinson's disease patients. Aging (Albany NY). 2015;**7** (12):1130–1142

[28] Levine ME, et al. Epigenetic age of the pre-frontal cortex is associated with neuritic plaques, amyloid load, and Alzheimer's disease related cognitive functioning. Aging (Albany NY). 2015;**7**:1198–1211

[29] Horvath S. DNA methylation age of human tissues and cell types. Genome Biology. 2013;**14**(10):3156

[30] Issa J-PJ, et al. Methylation of the oestrogen receptor CpG island links ageing and neoplasia in human colon. Nature Genetics. 1994; **7**:536–540

[31] Jung M, Pfeifer GP. Aging and DNA methylation. BMC Biology. 2015;**13**:7

[32] Bocklandt S, et al. Epigenetic predictor of age. PLoS One. 2011;**6**: e14821

[33] Horvath S. DNA methylation age of human tissues and cell types. Genome Biology. 2013;**14**(10):R115

[34] Calvanese V, et al. The role of epigenetics in aging and age-related diseases. Ageing Research Reviews. 2009;**8**:268–276

[35] Goel N, Karir P, Garg VK. Role of DNA methylation in human age prediction. Mechanisms of Ageing and Development. 2017;**166**:33–41

[36] Hannum G, et al. Genome-wide methylation profiles reveal quantitative views of human aging rates. Molecular Cell. 2013;**49**:359–367

[37] Horvath S, et al. Epigenetic clock for skin and blood cells applied to Hutchinson Gilford Progeria Syndrome and ex vivo studies. Aging (Albany NY). 2018;**10**:1758–1775

[38] Mondal T, et al. Characterization of the RNA content of chromatin. Genome Research. 2010;**20**:899–907

[39] Kaochar S, Tu BP. Gatekeepers of chromatin: Small metabolites elicit big changes in gene expression. Trends in Biochemical Sciences. 2012;**37**(11):477–483

[40] Liu B, Yip RK, Zhou Z. Chromatin remodeling, DNA damage repair and aging. Current Genomics. 2012;**13**(7):533–547

[41] McCulloch SD, Kunkel TA. The fidelity of DNA synthesis by eukaryotic replicative and translesion synthesis polymerases. Cell Research. 2008;**18**:148

[42] Saunders A, et al. Tracking FACT and the RNA Polymerase II Elongation Complex Through Chromatin in Vivo. Science. 2003;**301**:1094–1096

[43] Luger K, et al. Crystal structure of the nucleosome core particle at 2.8 Å resolution. Nature. 1997;**389**:251

[44] Mozzetta C, et al. Sound of silence: The properties and functions of repressive Lys methyltransferases. Nature Reviews Molecular Cell Biology. 2015;**16**:499

[45] Mersfelder EL, Parthun MR. The tale beyond the tail: Histone core domain modifications and the regulation of chromatin structure. Nucleic Acids Research. 2006;**34**(9):2653–2662

[46] Tropberger P, Schneider R. Scratching the (lateral) surface of chromatin regulation by histone modifications. Nature Structural & Molecular Biology. 2013;**20**:657

[47] Dong X, Weng Z. The correlation between histone modifications and gene expression. Epigenomics. 2013;**5**(2):113–116

[48] Strahl BD, Allis CD. The language of covalent histone modifications. Nature. 2000;**403**(6765):41–45

[49] Gomes AQ, Nolasco S, Soares H. Non-coding RNAs: Multi-tasking molecules in the cell. International Journal of Molecular Sciences. 2013;**14**(8):16010–16039

[50] Esteller M. Non-coding RNAs in human disease. Nature Reviews Genetics. 2011;**12**:861

[51] Collins LJ, Schönfeld B, Chen XS. Chapter 4 - The Epigenetics of non-coding RNA. In: Tollefsbol T, editor. Handbook of Epigenetics. San Diego: Academic Press; 2011. p. 49–61

[52] Palazzo AF, Lee ES. Non-coding RNA: What is functional and what is junk? Frontiers in Genetics. 2015;**6**:2

[53] Wilusz JE, Sunwoo H, Spector DL. Long noncoding RNAs: Functional surprises from the RNA world. Genes & Development. 2009;**23**(13):1494–1504

[54] Ulitsky I, Bartel DP. lincRNAs: Genomics, evolution, and mechanisms. Cell. 2013;**154**(1):26–46

[55] Kung JT, Colognori D, Lee JT. Long noncoding RNAs: Past, present, and future. Genetics. 2013;**193**(3):651–669

[56] Rutenberg-Schoenberg M, Sexton AN, Simon MD. The properties of long noncoding RNAs that regulate chromatin. Annual Review of Genomics and Human Genetics. 2016;**17**(1):69–94

[57] Roberts TC. The microRNA Machinery, in microRNA: Basic Science: From Molecular Biology to Clinical Practice. Santulli G, Editor. Cham: Springer International Publishing; 2015. p. 15–30

[58] Cheung P, et al. Single-cell chromatin modification profiling reveals increased epigenetic variations with aging. Cell. 2018;**173**:1385–1397.e14

[59] Lu Q, et al. Demethylation of ITGAL (CD11a) regulatory sequences in systemic lupus erythematosus. Arthritis and Rheumatism. 2002;**46**:1282–1291

[60] Sawalha AH, et al. Defective T-cell ERK signaling induces interferon-regulated gene expression and overexpression of methylation-sensitive genes similar to lupus patients. Genes and Immunity. 2008;**9**:368–378

[61] Moosavi A, Ardekani AM. Role of epigenetics in Biology and human diseases. Iranian Biomedical Journal. 2016;**20**(5):246–258

[62] Egger G, Liang G, Aparicio A, et al. Epigenetics in human disease and prospects for epigenetic therapy. Nature. 2024;**429**:457–463

[63] Robertson KD. DNA methylation, methyltransferase, and cancer. Oncogene. 2001;**20**(24):3139–3155

[64] Liu S, Sharma U. Sperm RNA payload: Implications for intergenerational epigenetic inheritance. International Journal of Molecular Sciences. 2023;**24**(6):5889

[65] Jimenez-Chillaron JC, et al. Transgenerational epigenetic inheritance of diabetes risk as a consequence of early nutritional imbalances. Proceedings of the Nutrition Society. 2016;**75**:78–89

[66] Legoff L, et al. Transgenerational inheritance of environmentally induced epigenetic alterations during mammalian development. Epigenomes. 2019;**8**:1559

Chapter 6

Key Role of Histone Modifiers in Coupling Metabolism to Epigenetic Reprogramming Associated with "Trained Immunity" in Innate Immune Cells

Anant Jaiswal and Timothy Osborne

Abstract

Trained immunity refers to an acquired immune memory component of innate immune cells, which respond quickly and robustly to secondary infections. It's very different from adaptive immunity since it is nonselective and doesn't depend on gene rearrangement; instead, epigenetic reprogramming acts as a foundation for trained immunity. Microbial or metabolic stimuli lead to epigenetic modifications that are influenced by metabolic pathway flux, resulting in enhanced inflammatory responses upon the secondary challenge. This chapter explores how epigenetic modifiers act as a bridge between metabolic pathway mediators and the proinflammatory response through a "poised" transcriptional state of chromatin that helps maintain a long-lasting epigenetic memory, influencing gene expression programs of innate immunity. We describe how cellular metabolites are not limited to metabolic pathway regulation, but they also influence the development and maintenance of a primed epigenetic landscape and shape inflammatory responses. We also discuss how different histone marks regulate gene expression during trained immunity and how key metabolites have unique roles in the function of histone-modifiers, including a direct role of metabolites in their enzymatic activity. We highlight major trained immunity inducers and how altered metabolite levels and epigenetic reprogramming integrate peripheral and central trained immunity for a long-lasting response. Finally, we analyze the pathological significance of trained immunity and how its protective capabilities against certain infections can result in maladaptive inflammation, particularly relevant to metabolic diseases, and we propose that a better understanding of epigenetic and metabolic networks during trained immunity has therapeutic potential relevant to immunological and metabolic disorders.

Keywords: trained immunity, histone modification, metabolism, epigenetics, innate immune cells

1. Introduction

Since the discovery of the human immune system, the notion of immunological memory has been reserved for the adaptive immune system, and less focus has been placed on innate immune branches. Antigen-driven gene rearrangements drive stable and irreversible adaptive memory in T and B lymphocytes, forming the basis of long-term antigen-specific memory that protects against pathogenic infection or the protective response elicited through vaccination [1, 2]. However, recent findings, especially over the past decade, have primarily challenged the concept that memory is unique to the adaptive system and suggest that innate immune cells can also develop a memory-like state by utilizing epigenetic reprogramming to adapt to pathogen exposure in a less-specific fashion. This has been referred to as "trained immunity" or "innate immune memory" [3].

Trained immunity refers to a state where innate immune cells, such as monocytes, macrophages, or neutrophils, exhibit an altered secondary immune response after a primary exposure to the same or different pathogenic, sterile-metabolic, or inflammatory stimuli [4, 5]. The altered secondary response is dependent on the nature and quantitative features of the secondary trigger, which results in enhancement or inhibition of the corresponding inflammatory response [6]. A key difference, relative to adaptive immunity, is that innate immune memory is not specific to a given pathogen, yet it shows a very robust secondary response that can last for several days and, in some documented cases, for months [7].

As mentioned above, adaptive immune memory develops by somatic recombination and clonal expansion of T and B cells [8], while trained immunity relies on epigenetic reprogramming of innate immune cells, in which modifications of histones and alterations to chromatin structure take place and are inherited through a limited number of cell divisions, without affecting underlying DNA sequences [9]. These epigenetic reprogramming events involve a coded series of "histone modifications," which add small molecular tags such as methyl, acetyl, or lactyl moieties to key histone amino acid side chains that affect chromatin openness or compactness, thereby influencing transcription of neighboring genes [6, 10, 11] These can occur through simple enzyme-dependent modification or through more dynamic "histone remodeling" complexes, which result in ATP-dependent restructuring of nucleosome positions through large multisubunit remodeling machines such as the SWI/SNF complex [12].

In trained immunity, these alterations lead to changes in the expression of proinflammatory and metabolic regulators and allow innate immune cells to "remember" their previous encounters with different inflammatory stimuli [6]. These trained cells show a stronger and more rapid response after exposure to a second signal, which can be unrelated to the primary stimulus. For instance, the addition of β-glucan (a fungal cell wall component) to cultured monocytes results in a secondary response, leading to enhanced proinflammatory cytokine production of TNF-α and IL-6 following secondary stimulation with a different pathogenic signal days later [6, 10, 13].

A fundamental discovery in the field of trained immunity over the last decade is the central role of metabolic intermediates in the regulation of epigenetic reprogramming in innate immune cells, which provides an intimate connection between cellular metabolism and innate immune training [10, 14, 15]. In this respect, changes in essential cellular metabolic pathway flux, such as a shift from oxidative phosphorylation to aerobic glycolysis, changes in the flux of the tricarboxylic acid (TCA) cycle, as well as glutaminolysis, have been shown to drive the epigenetic pathways affecting trained immunity [15]. This is at least partly because key intermediate metabolites produced in these pathways, like acetyl-CoA, fumarate, α-ketoglutarate (α-KG), or S-adenosylmethionine (SAM), function as essential cofactors and substrates for several histone modifiers. For example, SAM acts as a substrate for histone methyltransferases (HMTs) [16], acetyl-CoA is essential for histone acetyltransferase (HAT) function [17], and the TCA cycle intermediate α-KG is a crucial allosteric regulator of histone demethylases (KDMs) [18]. Consequently, these upstream metabolic regulators of chromatin modification connect a variety of immune triggers related to pathogens or metabolic stresses with trained immunity [19].

Histone-modifiers act as a bridge between metabolism and epigenetic remodeling during trained immunity. We discuss the role of histone-modifying enzymes like methyltransferases, acetyltransferases, deacetylases (HDACs), as well as histone demethylases (KDMs), and their regulation by metabolic intermediates during innate immune reprogramming. Understanding the connection between metabolic and epigenetic regulation will extend our knowledge of innate immune training and provide new therapeutic options centered around innate immunity to respond to pathogens, as well as chronic inflammatory disorders that perturb central metabolic flux and vice versa.

2. General concept of trained immunity

2.1 Trained immunity inducers

Sterile and pathogenic triggers that have been described in the literature include "Western diet" feeding-associated systemic inflammatory mediators like oxidized low-density lipoprotein (OxLDL) [9, 20], as well as damage-associated molecular patterns (DAMPs), including the following.

- **β-glucan:** a type of β-D-glucose polysaccharide that is present in the fungal cell wall and has been shown to induce immune training of innate immune cells through interaction with its cell surface receptor, Dectin-1 [21].

- **Bacillus Calmette-Guérin (BCG) vaccine:** clinical and mechanistic studies have shown that the BCG vaccine not only protects against *Mycobacterium tuberculosis* infection in an antigen-specific manner but also results in a broad pattern of defense against other unrelated infections from pathogens like *Candida albicans* or *influenza* [22]. In fact, this memory response from innate immune cells, achieved through BCG-induced trained immunity, has been substantial in decreasing overall mortality in vaccinated children [23].

2.2 Key cell types involved in trained immunity

Emerging research shows that several other nonimmune cell types, like epidermal stem cells, stromal or hematopoietic stem cells, also possess trained immunity features [24, 25]; however, the most well-studied cells are:

- **Neutrophils:** growing research suggests an indirect role of neutrophils during trained immunity, which is modulated by bone marrow progenitor reprogramming [26]. Immune training promotes granulopoiesis in bone marrow via increased proinflammatory mediators. This results in an enhanced number of granulocyte populations that further substantiate the inflammatory response against secondary infections [9].

- **Monocytes and macrophages:** these are the major cell types involved in trained immunity and have been extensively studied. In addition to their longer lifespan relative to neutrophils, they are also fundamental regulators of trained immunity [27]. For instance, primary exposure of monocytes to an immune trigger leads to robust epigenetic, transcriptional, and metabolic reprogramming, as well as their differentiation into macrophages [28]. These changes can endure for a long time, even for months, and the secondary exposure results in enhanced cytokine production and, in some cases at least, increased phagocytic activity.

Importantly, the mechanism for trained immunity induction can be highly specific to a particular cell type, and changes in the chromatin states and metabolic reprogramming can vary depending on the availability and identity of the key effector-metabolites responsible for the training response, as well as the life span and function of the trained cell. For example, monocyte/macrophage training may depend more on enhancer-promoter remodeling, while trained immunity in neutrophils relies mainly on the epigenetic remodeling of their progenitor populations, as described above [26–28].

2.3 Epigenetic reprogramming acts as the foundation of trained immunity

Epigenetic modifications are the mechanistic foundation of trained immunity; in fact, it is also known as "**epigenetic memory.**" These long-lasting modifications act as a bookmark and "prime" the innate immune cells for secondary responses. Some examples include:

- **Chromatin accessibility:** nucleosome remodeling or positioning in response to a wide variety of posttranslational modifications of histone tails leads to changes in chromatin structure that increase the accessibility of key transcription factors, such as NF-κB or AP-1, to their respective binding sites in chromatin, contributing to the enhanced proinflammatory response [12].

- **Histone modifications:** specific histone modifications within the promoters or enhancers of proinflammatory genes result in prompt activation after secondary exposure to an additional stimulus. For instance, the transfer of methyl groups to generate histone H3 lysine 4 di/trimethylation (H3K4me2/3) or acetyl groups to generate histone H3 lysine 27 acetylation (H3K27ac) at promoters and enhancers of Tnf/Il6 genes during training facilitates rapid gene transcription following secondary challenge [6, 10].

- **Epigenetic-metabolic crosstalk:** cellular metabolites such as lactate, acetyl-CoA, succinate, or fumarate also act as substrates for histone-modifying enzymes [11, 14, 16], and changes in their production by trained immunity inducers can directly affect the activity of the respective enzymes and thus influence the epigenetic landscape that shapes the proinflammatory signature of the affected cells [19].

Simultaneously, these epigenetic modifications, which are influenced by trained immunity inducers, also create a "poised" state for subsequent transcriptional activation by additional proinflammatory mediators, which leads to a sustained memory of an adaptable proinflammatory response.

3. Histone regulators in epigenetic reprogramming

Trained immunity can be divided into three phases: a **"priming"** or **"training phase,"** which includes an extensive epigenetic reprogramming in response to engagement with an initial triggering compound that results in increased/decreased chromatin accessibility through a specific combination of individual histone modifications. This is followed by a **"memory phase,"** represented by a "poised" or "bivalent" chromatin state, where key response genes are primed for the third **"reactivation phase,"** which is the rapid and robust transcriptional activation of genes in response to the secondary challenge.

Histone-modifying enzymes serve as an **"interpreter"** of the key training events that convert the pathogenic or sterile signals into long-lasting chromatin changes to regulate the trained immunity response [6, 10]. Some major well-known histone modifiers, as well as metabolic mediators that affect induced histone marks and regulate trained immunity, are:

3.1 Histone methyltransferases (HMTs)

HMTs are the enzymes that directly methylate the histones by using SAM as a donor for the methyl group during the enzymatic reaction [16]. Regulation of gene transcription by methylation is highly selective relative to the exact type and location of the methyl group on the amino acid side chain of the affected histone residue. For instance, tri-methylation at the lysine 4 position of histone (H3K4me3) in gene promoters has been associated with active transcription, while di and trimethylation at the lysine 9 position (H3K9me2/3) is associated most often with transcriptional repression [29, 30].

There are several lysine methyltransferases (KMTs) which catalyze site-specific methylation. For instance, SET7 increases H3K4 methylation levels at promoters of proinflammatory cytokines like TNF and IL-6 during innate immune training, which results in rapid activation of transcription of these genes after secondary exposure in the two-step innate training response [31]. In contrast, other KMTs like G9a, which is a member of the SET domain-containing lysine methyltransferase family, increase H3K9me2 levels at the promoters of these same cytokine genes, which are responsible for the negative regulation of trained immunity [32]. There are several other KMTs from the same SET family, such as SETDB1/2, that potentially have a role in trained immunity regulation and require further investigation.

In addition to H3K9 methylation, other lysine modifications, such as H3K27me3 have also been associated with transcriptional repression. For example, catalytic subunits of large protein complexes, such as EZH2 from the polycomb repressive complex 2 (PRC2), promote transcriptional repression by increasing H3K27me3 methylation levels on genes required for myeloid cell differentiation [33].

The opposing actions of the different functioning HMTs maintain a coordinated cascade of transcription activation and repression in the immune cells during the trained immunity response, providing a highly selective and finely tuned overall response.

3.2 Histone lysine demethylases (KDMs)

Histone methylation is balanced with demethylation to provide an additional layer of fine-tuning to the overall regulatory process. These lysine demethylases, or KDMs, also precisely remove methyl groups from specific histone positions. For instance, KDM4 (JMJD2) removes methyl groups from repressive marks such as H3K9me3 and H3K36me3, leading to transcriptional activation [34], while the KDM5 (JARID1) family members remove methyl groups from active histone marks like H3K4me3 and their action results in reduced gene transcription [14]. Another demethylase, KDM6B (JMJD3), removes the repressive H3K27me3 marks, leading to the induction of proinflammatory gene expression during immune training [35].

Importantly, these KDMs utilize a specific intermediate metabolite of the TCA cycle, α-ketoglutarate (α-KG), as a crucial allosteric cofactor [18], and they are also negatively regulated by fumarate and succinate accumulation during immune training [14]. This further highlights the integrated connection between metabolic flux and epigenetic modification during trained immunity.

3.3 Histone acetyltransferases (HATs)

HATs represent a transfer of the acetate moiety from acetyl-CoA to the epsilon amino group of key lysine residues of histone tails [17]. These modifications are thought to mask the overall positive charge of the histone proteins and decrease their association with negatively charged DNA within nucleosomes, which further leads to the opening of chromatin and transcriptional activation. Among these HATs, CBP and p300 play a pivotal role in trained immunity [19]. Because HATs use acetyl-CoA, which is a central node in cellular energy metabolism, as a source of the acetyl group, this provides another key connection between metabolic flux and the epigenetic regulation of gene expression [36]. Trained immunity inducers also trigger metabolic reprogramming by shifting innate immune cell energy metabolism away from oxidative phosphorylation to favor increased glycolysis [6]. This further emphasizes the connection between metabolic flux through acetyl-CoA and innate immune memory.

3.4 Histone deacetylases (HDACs)

Similar to KDMs that remove methyl groups from histones, there are several HDACs that remove acetyl groups from histones, thereby returning the overall positive charge to the histone protein, which favors chromatin condensation [37]. There are two main subclasses of HDACs: classical HDACs (class I and II), which use stably bound zinc as a cofactor for their enzyme activity. These include HDAC1/2, and, along with NuRD and Sin3A, they form a transcriptional repressor complex, leading to deacetylation of histones and inhibition of gene transcription in most cases [38]. The other subclass includes the sirtuins (class III HDACs). Like KDMs, the activity of the sirtuin family (Sirtuin 1-5) is also dependent on cellular metabolism and is regulated by the redox-sensitive energy sensor NAD + [39]. Therefore, sirtuin-dependent metabolic reprogramming of cellular pathways connects the NAD +/NADH ratio and cellular redox balance to trained immunity [40]. In fact, HDACs can modulate the gene activation or repression state of chromatin in a context-specific manner; for example, in some cases, HDACs negatively regulate certain inflammatory genes, while in other settings, they can modulate immune response pathways by preventing the restoration of repressive marks in the genome [41].

3.5 Histone lactylation

Modification of histones through lactylation is also involved in trained immunity and provides another compelling link between metabolism and epigenetic regulation.

A recent study has shown that monocytes use lactate preferentially as a TCA cycle substrate instead of glucose during trained immunity, and this study also demonstrated that histone lactylation is an essential modification for the trained immunity response to pathogenic infection [42].

Another recent study highlighted the importance of histone lactylation during BCG-induced trained immunity, where an increase in lactate accumulation resulted in increased lactylation of histone H3 at lysine 18 (H3K18la) in the regulatory regions of key cytokine and chemokine genes [11]. This mark is associated with active transcription and enhances the cytokine response against secondary challenges. Blocking lactate production or lactylation directly using inhibitors resulted in a decrease in the trained immunity response in innate immune cells. This identifies H3K18la as another histone mark regulating the epigenetic landscape of chromatin during trained immunity [11].

Although histone lactylation and histone acetylation have similar effects on gene activation and are both related to lysine modification, their different metabolic substrates influence how they selectively modulate trained immunity. Histone acetylation, which is an early event in training, is influenced primarily by acetyl-CoA levels and occurs during the initial phase of the training response, whereas lactylation is directed by the increase in lactate, which accumulates after the metabolic shift to aerobic glycolysis has occurred, and is thought to primarily influence later steps in the training response.

Overall, the integrated system of histone modifications influences complex metabolic-epigenetic regulatory networks, relaying changes in cellular metabolic balance, resulting in a stable, poised, or primed state of chromatin structure, which increases the capacity of immune cells to more effectively combat secondary infections.

4. Metabolic regulation of the epigenetic landscape in trained immunity

This section discusses how cellular metabolic mediators and molecular pathways regulate the epigenetic landscape, from inducing an immune memory to strengthening a long-lasting and heritable program of trained immunity in the innate immune system.

4.1 Cellular metabolic rewiring is an initial event in trained immunity

Metabolic reprogramming of immune cells is the initial step after interaction with trained immunity inducers such as BCG, β-glucan, or cellular metabolites. For instance, the classic immune trigger β-glucan interacts with its cell surface receptor dectin-1 and activates a downstream signaling cascade, including the Akt-mTOR pathway, which further increases the shift from oxidative phosphorylation toward glycolysis [10]. This metabolic shift accelerates downstream metabolic cycle flux and alters the production of metabolites such as lactate, pyruvate, acetyl-CoA, and fumarate, which are not only essential metabolic pathway intermediates, but also serve as regulators and/or cosubstrates for key epigenetic enzymes [10, 14, 36].

Therefore, this early metabolic reprogramming, on the one hand, changes the metabolic machinery to favor the immediate energy requirements of the cell during immune activation, while, on the other hand, serves as a signal to induce and sustain a long-term memory response by enhancing epigenetic remodeling of chromatin in innate immune cells.

4.2 Cellular metabolites directly regulate epigenetic modifiers

Several metabolites are directly involved in the function of epigenetic enzymes, such as.

- **Lactate:** increased in the anaerobic glycolytic shift during immune activation and leads to cellular accumulation of lactate [42], which is also directed to the nucleus, where it is transferred to histone H3K18 on histones within the regulatory region of key proinflammatory cytokine genes to promote trained immunity [11].

- **Acetyl-CoA:** high pyruvate levels from glycolysis or from lactate fuel mitochondria to produce Acetyl-CoA [36]. The increased Acetyl-CoA serves as an essential cofactor for the function of HATs like p300 or CBP, which further increases H3K27ac marks in the enhancer regions of proinflammatory genes, leading to chromatin unfolding, which makes these regulatory regions more accessible to the transcriptional machinery for gene activation [17].

- **NAD^+:** it is not only an important player in cellular redox balance but also acts as a cofactor for epigenetic enzymes like HDACs, particularly SIRT family deacetylases [39], thereby regulating the histone acetylation levels of the proinflammatory gene and its transcription in a context-specific manner [40].

- **SAM:** is produced during methionine metabolism [43], and it also acts as a universal carrier of methyl groups for almost all cellular methyl transferase reactions, including histone methylation through HMTs [16].

- **α-Ketoglutarate:** a key metabolite of the TCA cycle, also serves as an essential cofactor for histone demethylases like KDM4 and KDM5/6 [18]. Abundant α-KG levels help gene activation and repression via the removal of specific histone methylation marks, such as H3K4me3 via KDM5 [14], and H3K9me3 and H3K9me1 via KDM4 and KDM3A respectively [44].

Collectively, enhanced metabolic flux, along with abundant metabolite levels, fulfills higher cellular bioenergetic and redox requirements and relates these signals to the epigenetic landscape of chromatin through histone modification to provide long-lasting responses for the immune training phenotype (**Figure 1**).

5. Types of trained immunity and their integration in immune activation

Recent advancements in research studies have highlighted two main complementary categories of trained immunity, including

5.1 Peripheral trained immunity

Acts as a first line of defense against pathogens and involves several differentiated innate immune cells such as neutrophils, monocytes, or specific tissue macrophages, which are already present in circulation [45]. Immune inducers such as BCG, β-glucan, or OxLDL are all known to induce metabolic and epigenetic changes in these cells, leading to hyper-activation after a secondary stimulus [6, 10].

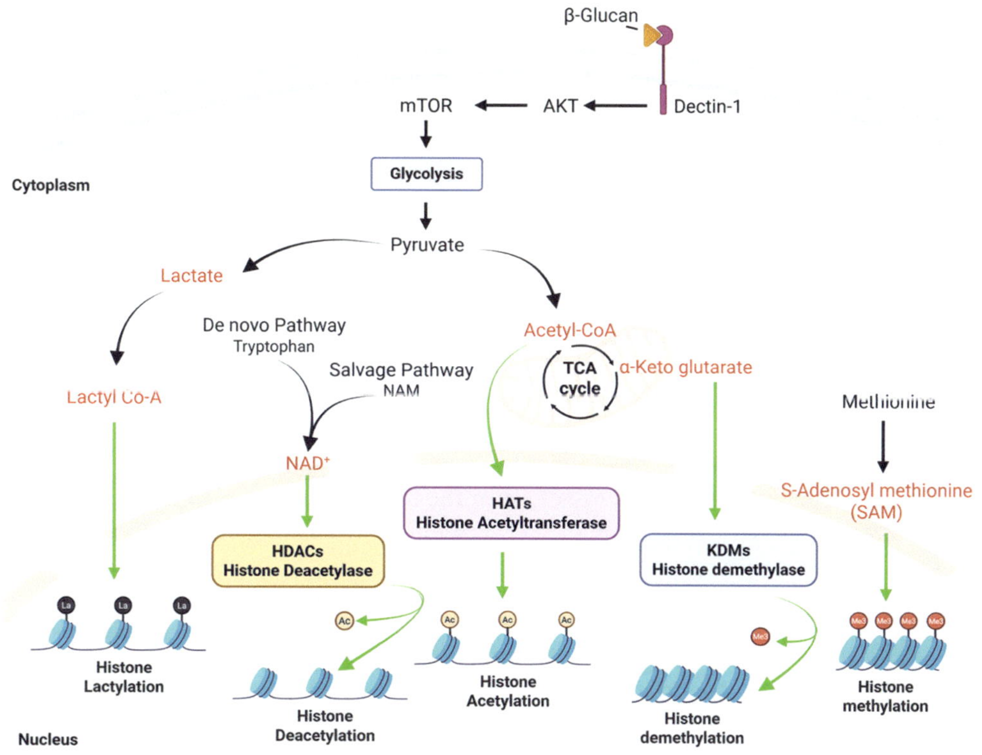

Figure 1.
Metabolic and epigenetic mediators' coordination during trained immunity. B-glucan-mediated induction of glycolysis via the AKT/mTOR pathway leads to high levels of, which directly participate in histone lactylation. Similarly, acetyl-CoA and α-ketoglutarate regulate histone acetyltransferase and histone demethylase function, respectively. Furthermore, S-adenosylmethionine and NAD + are involved in histone methylation and deacetylation. (Created by BioRender).

Despite their fast action toward these inducers, these relatively short-lived cells in the circulation limit their effectiveness for long-term immune memory reprogramming.

5.2 Central trained immunity

More recent studies have shown that metabolic and epigenetic signals interact, leading to innate immune training that is not limited to peripheral or circulating innate immune cells. Instead, it extends to the precursor bone marrow cells, resulting in a more prolonged memory of the initial triggering event [13, 26]. In this respect, trained immunity inducers also lead to changes in precursor and progenitor cells present in the bone marrow. For instance, BCG, β-glucan, or even western-diet signals have been shown to induce metabolic and epigenetic changes not only in peripheral innate immune cells but also in the bone marrow hematopoietic stem and progenitor cells (HSPCs) [9, 13]. These changes further increase precursor cell commitment toward the myeloid lineage and lead to a shift from lymphopoiesis to increased myelopoiesis. These epigenetic changes in the progenitor cells can persist through many rounds of cell division and lead to training effects that are maintained for prolonged periods, which complement the effects of trained immunity on the short-lived circulating innate immune cells [9].

Like peripheral immunity, central trained immunity also involves similar metabolic and epigenetic reprogramming, including altered Akt-mTOR signaling to modulate glycolysis and TCA cycle activity, resulting in similar histone modifications like acetylation or methylation in the bone marrow progenitors [13]. However, unlike peripheral effects, where cells are relatively short-lived, central trained immunity can persist for extended periods, lasting from weeks to months [46] (**Figure 2**).

Together, peripheral and central trained immunity shape an integrated and complementary network, utilizing similar metabolic and epigenetic reprogramming mechanisms, which provide the innate immune system with a robust memory to combat secondary exposure to pathogens in a faster and more effective manner.

6. Implications of trained immunity for disease pathology

The concept of a memory component for the innate immune system has reformed our fundamental view of the role of innate immune cell function in inflammatory and metabolic diseases.

The beneficial effect of immune training by enhancing the host defense capabilities against secondary pathogens has emerged as a crucial protective factor in several infectious diseases [13, 47]. For instance, the BCG vaccine has broad clinical effectiveness beyond the primary target of *Mycobacterium Tuberculosis* because it also induces broad protection against other viruses and bacteria through trained immune reprogramming of innate immune cells [23]. Through metabolic and epigenetic remodeling, it not only changes the transcriptional state of peripheral immune cells but also affects the progenitor immune population in the bone marrow to provide long-lasting, broad protection against unrelated pathogens [26, 46]. The immune training capabilities of BCG have also been shown to be effective in other immunotherapy settings, including cancer [48].

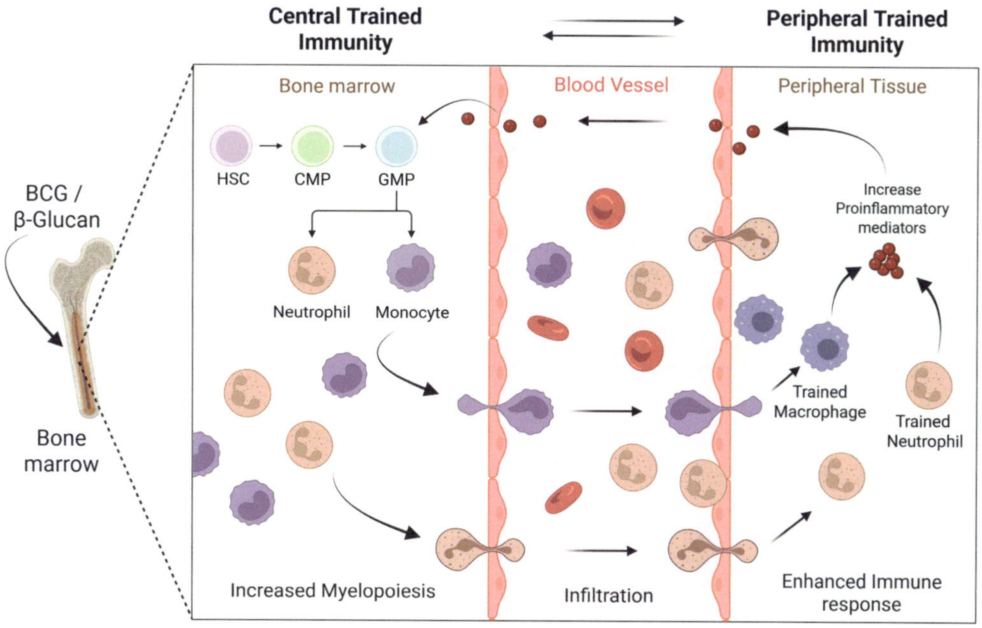

Figure 2.
Integrated network of central and peripheral trained immunity. Trained immunity triggers induce an increase in myelopoiesis in the bone marrow. This enhances HSC differentiation into granulocytes, leading to increased monocyte and neutrophil numbers and infiltration into peripheral tissues, which ultimately results in a hyper proinflammatory response. These proinflammatory mediators further communicate to the bone marrow precursor and progenitor cells to maintain a long-lasting trained immune response. (Created by BioRender).

In addition, β-glucan, a fungal cell wall component, has been shown to enhance innate immune activation via inducing epigenetic memory and provides resistance toward unrelated secondary bacterial or viral infections [13, 47]. Additionally, it also improves the anti-tumor capability of cells with an enhanced proinflammatory response [49].

Although trained immunity can be beneficial in increasing host defense against pathogens, prolonged activation of the proinflammatory response can be detrimental and may result in chronic inflammatory conditions, leading to sterile inflammation and metabolic diseases [26]. One of the best examples is atherosclerosis, where sterile proinflammatory molecules accumulate during the consumption of a high-fat “Western-diet” [9]. One such component is OxLDL, which serves as a potent proinflammatory and innate immune training signal [20]. The OxLDL-trained progenitors then carry this hyper-response to their descendant immune cells, such as monocytes and neutrophils, which contribute to the chronic inflammation associated with atherosclerosis progression [9].

Thus, overall, trained immunity can act as both friend and foe, where, on one hand, it contributes to the protection against certain pathogens by increasing host defense capabilities, while, on the other hand, the potentially associated dysfunctional or prolonged trained responses lead to chronic inflammatory conditions, resulting in metabolic diseases, including atherosclerosis. Therefore, it is likely that a better understanding of the underlying epigenetic and molecular mechanisms involved will offer new therapeutic opportunities for targeting both infectious and chronic inflammatory diseases for future generations.

7. Conclusion

Trained immunity has reshaped our understanding of how the host innate immune system limits pathogen damage. For decades, innate immune action against pathogens was considered to be a limited primary response, while more recent findings have established that innate immune cells can also learn from previous pathogenic encounters, providing a form of "molecular memory" that can be recalled to mount a more robust response to secondary challenges. Importantly, the memory mechanism does not originate from permanent gene rearrangement events like in adaptive immunity, but it is based on a combinatorial pattern of epigenomic responses that establish a primed chromatin template to respond to secondary challenges and provide a long-lasting form of immune memory.

Although this strong immune response is often protective against pathogens, it can also result in uncontrolled hyper-responses that contribute to chronic inflammation, also known as "maladaptive inflammation," as in the case of metabolic disorders. Therefore, these molecular and epigenetic mechanisms are very context-dependent, where the same mechanism increases protection against pathogens, but in other settings, they can also promote disease pathology.

Overall, a better understanding of how metabolites and histone modifiers coordinate together with additional factors that regulate their function will provide better precision therapeutic opportunities for selectively targeting and treating a wide range of immunological and metabolic disorders.

Acknowledgments

The authors were supported in part by a grant from the NIH to TO (DK124343). We also thank the anonymous reviewer for helpful comments on the manuscript.

Conflict of Interest

The authors declare no conflicts of interest.

Author details

Anant Jaiswal[1,2] and Timothy Osborne[1,2]*

1 Division of Diabetes, Endocrinology, and Metabolism, Departments of Medicine, Pediatrics, Physiology, Pharmacology, and Therapeutics, Johns Hopkins University School of Medicine, Saint Petersburg, FL, United States of America

2 Institute for Fundamental Biomedical Research, Johns Hopkins All Children's Hospital, Saint Petersburg, FL, United States of America

*Address all correspondence to: tosborn9@jh.edu

References

[1] Gattinoni L, Speiser DE, Lichterfeld M, Bonini C. T memory stem cells in health and disease. Nature Medicine. 2017;**23**(1):18–27

[2] Inoue T, Kurosaki T. Memory B cells. Nature Reviews Immunology. 2024;**24**(1):5–17

[3] Ochando J, Mulder WJM, Madsen JC, Netea MG, Duivenvoorden R. Trained immunity-basic concepts and contributions to immunopathology. Nature Reviews Nephrology. 2023;**19**(1):23–37

[4] Leentjens J, Bekkering S, Joosten LAB, Netea MG, Burgner DP, Riksen NP. Trained innate immunity as a novel mechanism linking infection and the development of atherosclerosis. Circulation Research. 2018;**122**(5):664–669

[5] Kalafati L, Hatzioannou A, Hajishengallis G, Chavakis T. The role of neutrophils in trained immunity. Immunological Reviews. 2023;**314**(1):142–157

[6] Novakovic B, Habibi E, Wang SY, Arts RJW, Davar R, Megchelenbrink W, et al. Beta-Glucan Reverses the epigenetic state of LPS-induced immunological tolerance. Cell 2016;**167**:1354–68e14

[7] Kleinnijenhuis J, Quintin J, Preijers F, Benn CS, Joosten LA, Jacobs C, et al. Long-lasting effects of BCG vaccination on both heterologous Th1/Th17 responses and innate trained immunity. Journal of Innate Immunity 2014;**6**:152–158

[8] Chi X, Li Y, Qiu X. V(D)J recombination, somatic hypermutation and class switch recombination of immunoglobulins: Mechanism and regulation. Immunology. 2020;**160**(3):233–247

[9] Christ A, Gunther P, Lauterbach MAR, Duewell P, Biswas D, Pelka K, et al. Western diet triggers NLRP3-dependent innate immune reprogramming. Cell 2018;**172**:162–75e14

[10] Bekkering S, Arts RJW, Novakovic B, Kourtzelis I, van der Heijden C, Li Y, et al. Metabolic induction of trained immunity through the mevalonate pathway. Cell 2018;**172**:135–46e9

[11] Ziogas A, Novakovic B, Ventriglia L, Galang N, Tran KA, Li W, et al. Long-term histone lactylation connects metabolic and epigenetic rewiring in innate immune memory. Cell 2025;**188**:2992–3012e16

[12] Chen Y, Liang R, Li Y, Jiang L, Ma D, Luo Q, et al. Chromatin accessibility: Biological functions, molecular mechanisms and therapeutic application. Signal Transduction and Targeted Therapy 2024;**9**:340

[13] Moorlag S, Khan N, Novakovic B, Kaufmann E, Jansen T, van Crevel R, et al. Beta-glucan induces protective trained immunity against Mycobacterium tuberculosis Infection: A key role for IL-1. Cell Reports 2020;**31**:107634

[14] Arts RJ, Novakovic B, Ter Horst R, Carvalho A, Bekkering S, Lachmandas E, et al. Glutaminolysis and Fumarate Accumulation Integrate Immunometabolic and Epigenetic

Programs in Trained Immunity. Cell Metabolism 2016;**24**:807–819

[15] Arts RJ, Joosten LA, Netea MG. Immunometabolic circuits in trained immunity. Seminars in Immunology. 2016;**28**(5):425–430

[16] Abdelraheem E, Thair B, Varela RF, Jockmann E, Popadic D, Hailes HC, et al. Methyltransferases: Functions and Applications. Chembiochem 2022;**23**:e202200212

[17] Shi L, Tu BP. Acetyl-CoA and the regulation of metabolism: Mechanisms and consequences. Current Opinion in Cell Biology. 2015;**33**:125–131

[18] Xiao M, Yang H, Xu W, Ma S, Lin H, Zhu H, et al. Inhibition of alpha-KG-dependent histone and DNA demethylases by fumarate and succinate that are accumulated in mutations of FH and SDH tumor suppressors. Genes & Development 2012;**26**:1326–1338

[19] Fanucchi S, Dominguez-Andres J, Joosten LAB, Netea MG, Mhlanga MM. The intersection of epigenetics and metabolism in trained immunity. Immunity. 2021;**54**(1):32–43

[20] Scarpa A, Jung Y, Hamid A, Matzaraki V, More T, Heinz A, et al. Trained immunity induced by oxidized low-density lipoprotein is dependent on glutaminolysis. The FASEB Journal 2025;**39**:e70774

[21] Mata-Martinez P, Bergon-Gutierrez M, Del Fresno C. Dectin-1 signaling update: New perspectives for trained immunity. Frontiers in Immunology. 2022;**13**:812148

[22] Covian C, Fernandez-Fierro A, Retamal-Diaz A, Diaz FE, Vasquez AE, Lay MK, et al. BCG-induced cross-protection and development of trained immunity: Implication for vaccine design. Frontiers in Immunology. 2019;**10**:2806

[23] de Castro MJ, Pardo-Seco J, Martinon-Torres F. Nonspecific (Heterologous) protection of neonatal BCG vaccination against hospitalization due to respiratory infection and sepsis. Clinical Infectious Diseases. 2015;**60**(11):1611–1619

[24] Naik S, Larsen SB, Gomez NC, Alaverdyan K, Sendoel A, Yuan S, et al. Inflammatory memory sensitizes skin epithelial stem cells to tissue damage. Nature 2017;**550**:475–480

[25] Kaufmann E, Sanz J, Dunn JL, Khan N, Mendonca LE, Pacis A, et al. BCG educates hematopoietic stem cells to generate protective innate immunity against tuberculosis. Cell 2018;**172**:176–90e19

[26] Lavillegrand JR, Al-Rifai R, Thietart S, Guyon T, Vandestienne M, Cohen R, et al. Alternating high-fat diet enhances atherosclerosis by neutrophil reprogramming. Nature. 2024;**634**(8033):447–456

[27] Damani-Yokota P, Khanna KM. Innate immune memory: The evolving role of macrophages in therapy. Elife. 2025;**14**:e108276

[28] Saeed S, Quintin J, Kerstens HH, Rao NA, Aghajanirefah A, Matarese F, et al. Epigenetic programming of monocyte-to-macrophage differentiation and trained innate immunity. Science. 2014;**345**(6204):1251086

[29] Yu H, Lesch BJ. Functional roles of H3K4 methylation in transcriptional

regulation. Molecular and Cellular Biology. 2024;**44**(11):505–515

[30] Pourreau-Schneider N. In vitro growth promotion of rat leukemia L5222 cells in the presence of 2-mercaptoethanol. Journal of the National Cancer Institute. 1975;**55**(6):1467–1470

[31] Keating ST, Groh L, van der Heijden C, Rodriguez H, Dos Santos JC, Fanucchi S, et al. The set7 lysine methyltransferase regulates plasticity in oxidative phosphorylation necessary for trained immunity induced by beta-glucan. Cell Reports. 2020;**31**:107548

[32] Mourits VP, van Puffelen JH, Novakovic B, Bruno M, Ferreira AV, Arts RJ, et al. Lysine methyltransferase G9a is an important modulator of trained immunity. Clinical & Translational Immunology 2021;**10**:e1253

[33] Xie H, Xu J, Hsu JH, Nguyen M, Fujiwara Y, Peng C, et al. Polycomb repressive complex 2 regulates normal hematopoietic stem cell function in a developmental-stage-specific manner. Cell Stem Cell 2014;**14**:68–80

[34] Moorlag S, Matzaraki V, van Puffelen JH, van der Heijden C, Keating S, Groh L, et al. An integrative genomics approach identifies KDM4 as a modulator of trained immunity. European Journal of Immunology 2022;**52**:431–446

[35] Han L, Li L, Yao L, Bu H, Tian Y, Li Q, et al. H3K27me3 modulates trained immunity of monocytes in HDM-allergic diseases. Frontiers in Immunology. 2025;**16**:1572796

[36] Pietrocola F, Galluzzi L, Bravo-San Pedro JM, Madeo F, Kroemer G. Acetyl coenzyme A: A central metabolite and second messenger. Cell Metabolism. 2015;**21**(6):805–821

[37] Park SY, Kim JS. A short guide to histone deacetylases including recent progress on class II enzymes. Experimental & Molecular Medicine. 2020;**52**(2):204–212

[38] Adams GE, Chandru A, Cowley SM. Co-repressor, co-activator and general transcription factor: The many faces of the Sin3 histone deacetylase (HDAC) complex. Biochemical Journal. 2018;**475**(24):3921–3932

[39] Sauve AA, Youn DY. Sirtuins: NAD(+)-dependent deacetylase mechanism and regulation. Current Opinion in Chemical Biology. 2012;**16**(5–6):535–543

[40] Xu W, Guo Z, Xu T, Chen J, Chen L, Xu W. Reversing inflammatory diseases via trained immunity: Mechanisms, challenges, and prospects. Frontiers in Immunology. 2025;**16**:1666233

[41] Mohammadi A, Sharifi A, Pourpaknia R, Mohammadian S, Sahebkar A. Manipulating macrophage polarization and function using classical HDAC inhibitors: Implications for autoimmunity and inflammation. Critical Reviews in Oncology/Hematology. 2018;**128**:1–18

[42] Cai H, Chen X, Liu Y, Chen Y, Zhong G, Chen X, et al. Lactate activates trained immunity by fueling the tricarboxylic acid cycle and regulating histone lactylation. Nature Communications 2025;**16**:3230

[43] Igarashi K, Katoh Y. Metabolic aspects of epigenome: Coupling of S-adenosylmethionine synthesis and gene regulation on chromatin by SAMIT

module. Subcellular Biochemistry. 2013;**61**:105–118

[44] Qu L, Yin T, Zhao Y, Lv W, Liu Z, Chen C, et al. Histone demethylases in the regulation of immunity and inflammation. Cell Death Discovery 2023;**9**:188

[45] You M, Chen L, Zhang D, Zhao P, Chen Z, Qin EQ, et al. Single-cell epigenomic landscape of peripheral immune cells reveals establishment of trained immunity in individuals convalescing from COVID-19. Nature Cell Biology 2021;**23**:620–630

[46] Sun SJ, Aguirre-Gamboa R, de Bree LCJ, Sanz J, Dumaine A, van der Velden W, et al. BCG vaccination alters the epigenetic landscape of progenitor cells in human bone marrow to influence innate immune responses. Immunity 2024;**57**:2095–107e8

[47] Khan N, Tran KA, Chevre R, Locher V, Richter M, Sun S, et al. beta-Glucan reprograms neutrophils to promote disease tolerance against influenza A virus. Nature Immunology 2025;**26**:174–187

[48] van Puffelen JH, Keating ST, Oosterwijk E, van der Heijden AG, Netea MG, Joosten LAB, et al. Trained immunity as a molecular mechanism for BCG immunotherapy in bladder cancer. Nature Reviews Urology 2020;**17**: 513–525

[49] Kalafati L, Kourtzelis I, Schulte-Schrepping J, Li X, Hatzioannou A, Grinenko T, et al. Innate immune training of granulopoiesis promotes anti-tumor activity. Cell 2020;**183**: 771–85e12